CEP: Insights from Data to Action

Jacob

Table of Contents

Chapter 1

Introduction

Complex Event Processing (CEP) combines simple events to infer complex events that signify a specific situation of interest (*SOI*). In general terms, a complex event is an event that summarizes, represents, or denotes a set of other simple events [3, 4].

The core feature of CEP technology is in handling large volumes of information in near real-time, hence, it is an appropriate paradigm to enforce the concept of real-time situational awareness by providing the ability to react to situations of interest as soon as they occur [5–7].

In general terms, situation awareness is the perception of environmental elements concerning time or space, the comprehension of their meaning, and the projection of their future status [8].

From the situational awareness perspective, acquiring the state of knowledge by a identifying a particular *SOI* that implies an opportunity or threat requires gathering sufficient facts within a specified time and space (e.g., geospatial boundaries) [7, 9]. Hence, to apply this concept within the CEP domain, the required facts must be inferred, analyzed, and delivered in near real-time [10–12]. To achieve this, CEP systems rely on detection rules defined on complex pattern queries to filter, correlate, and aggregate events in near real-time.

The main issue related to identifying situations in unbounded event streams is the continuous need to learn and update CEP detection rules to adapt to emerging situations more efficiently [11, 13, 14]. Such challenge exists because CEP based systems obtain its input data as a continuous feed (real-time event stream) from sources such as sensors, RFIDs, etc. [10].

Any data stream mining approach used for CEP systems must have a high degree of flexibility to adapt to the changes in situations of interest in near real-time [14–17].

Several research efforts have pointed out the challenge of applying data mining techniques to learn CEP detection rules in real-time streaming environments especially with limited time available to infer such rules from the event stream [18, 19]. Existing rule learning approaches either focus on mining CEP rules from historical instances of events [20–23] or external domain knowledge [13, 14]. Other attempts have been proposed to incorporate domain knowledge with historical events in the rule learning process such as the *SCEPter* framework [24].

Despite partial success in automating CEP rule generation process either by fine-tuning existing CEP rules [23, 25], or by generating entirely new CEP rules from historical events [20, 21], dealing with the increased number of rules due to the unbounded nature of event stream is still a challenging task. Specifically, an increased number of CEP detection rules lead to more complex detection patterns, which makes the targeted situation more complicated to comprehend and handle [12, 20, 26]. Furthermore, such an increase will lead to higher levels of computational complexity [27]. On the other hand, limiting the number of CEP rules might not generate enough rules that reflect the desired level of situation refinement.

Thus, tackling these challenges requires achieving an acceptable trade-off between situation complexity and the number of CEP rules needed to identify that situation; this leads us to the primary question related to our work:

How to gain the desired level of situation refinement while maintaining the minimum number of CEP rules?

1.1   Background of the Proposed Model

In our work in [10] and [28], we investigated various motivations and challenges that drive various methods toward situation refinement; as a result, we have identified the following key features that must be considered to build an efficient Situation Refinement Model.

1. **Context Awareness:** The rule-tuning process must be enriched with the Domain Knowledge to present situation of interest based on the user's preferred context.

2. **Compliance with Event Streaming Constraints:** The situation refinement process must be fulfilled in near real-time and in a single scan regardless of the size and the rate of the event stream.

3. **Cost-Based:** To achieve the highest gain with the minimum cost, the Situation Refinement Model must identify the best tradeoff concerning refinement cost and refinement gain.

4. **Scenario-Driven:** Since the real-world scenarios differ in their business needs; the **SRM** must take into account the unique refinement requirement of the targeted scenario.

In our work, we undertake the problem of situation refinement by considering these features combined. Section 1.1.1 provides a background of the key features of the proposed model.

1.1.1 Key Features of the Proposed Model: Background

1.1.1.1 Context-Aware Situation Refinement

The event stream sources, such as *RFID* readers or sensor data streams, can only produce a limited amount of data due to their limited memory and event streaming throughput and latency restrictions.

To compensate such a lack of information, it is essential to enrich primitive event data with domain knowledge to enhance situational awareness about the current state of the system. Specifically, associate the presence of detected situations to some environmental attributes residing in surrounding domain knowledge.

Fig. 1.1 shows how to use the proposed Situation-Refinemnt model to refine existing situation S in a car sales system. As shown in this example, the **SRM** uses the *"Country Of Origin"* knowledge attribute to provide several refined representations $(SOI_1,...SOI_n)$ for the original detected situation.

Using domain knowledge to enrich event data implies embedding the proper detection rules into the complex pattern to represent *SOI* according to the user's preferred context. Two challenges must be tackled to fulfill such a purpose: First, CEP detection rules must be enriched over time to synchronize with the current state of domain knowledge; and second, the enrichment process must be fulfilled in near real-time to cope with the continuous arrival of the event stream and to avoid data queuing.

In this work, we undertake this issue by enriching CEP detection rules with the current state (value) of the selected knowledge property represents the user's preferred context.

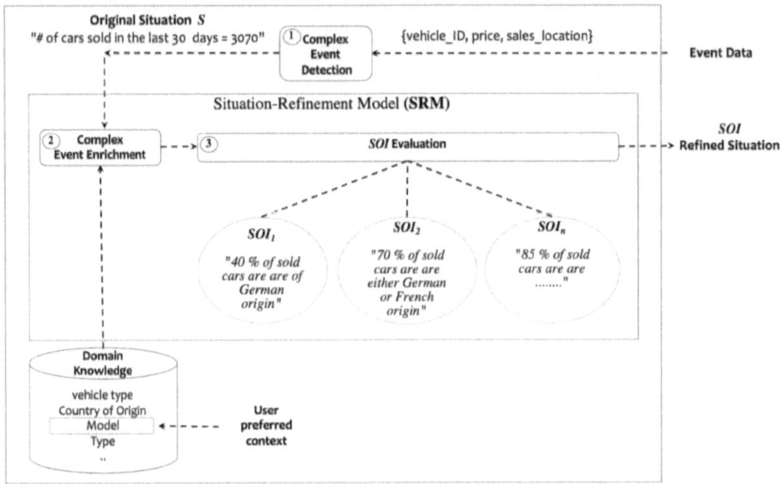

Figure 1.1: The Situation Refinement Model

1.1.1.2 Compliance with Event Streaming Constraints

Data mining over event streams has some challenging characteristics, such as the unbound size of the event stream, fast arrival rate, and inability to revisit previously arrived event data. Therefore, the following restrictions must be applied during the mining process in event streams:

1. Each data element of the stream should be examined at most once.

2. The memory usage during the mining process must be restricted regardless of the size of the event stream.

3. Each element in the stream should be processed in near real-time to avoid data queuing.

4. The mining outcome must be accessible instantly when the user requested.

In addition to these restrictions, two constraints must be imposed when using CEP systems to discover interest situations:

1. CEP systems aim at detecting emerging situations and respond to them in near real-time. These responses compose of a set of automated reactions triggered

as soon as a complex event is detected. The fulfillment of these responses requires sufficient resources to react within the event stream throughput limits. Therefore, any increase in the number of detected complex events might cause a delayed response and task queuing due to insufficient resources.

Most of the CEP systems tend to manage the increased number of complex events by enhancing the system scalability to accommodate more resources. Although such an approach is visible with systems that require the only automated type of response, in some scenarios, it is difficult to achieve a higher level of scalability, as the resources available are limited by nature or hard to replicate.

To tackle this issue, we propose the following contribution:

•**Contribution #1**: Using a custom support threshold to govern the level of situation refinement while maintaining the minimum number of CEP rules.

This contribution's significance is the ability to maintain a manageable number of complex events to match the available resources at any given time.

2. Due to the dynamically changing nature of the event streams over time, factors such as seasonal trends, the distribution of events in the stream, and the frequency of occurrence per time window must be considered to identify the targeted *SOI* accurately.

To tackle this issue, we present a Weighted Frequent Item Mining (*WFIM*) algorithm to re-evaluate and update the refinement summary over time.

• **Contribution #2**: Using the Relative Weight Factor to re-evaluate the mining outcome as the event stream evolves.

1.1.1.3 Cost-Based Situation Refinement

Refining CEP detection rules to achieve a higher level of situational awareness requires using an expressive pattern specification to identify emerging situations based on the reasoning over event streams and the available domain knowledge. Most of the CEP-tuning methods tend to increase the expressiveness of complex patterns to enhance and refine detected situations. The increased pattern expressiveness leads to two adverse outcomes: First, an increase in the system latency as a result of additional computation required to perform reasoning tasks, and second, more pattern complexity due to the increased number of CEP detection rules, which leads to more complex situation representation that is challenging to analyze and comprehend.

The increased pattern complexity negatively impacts the overall system performance concerning application throughput and latency requirements since event streaming constraints imply performing such process in a small time-window.

Hence, it is essential to consider refinement cost and gain when evaluating the overall performance of the situation refinement, in particular, assessing the cost of the additional pattern complexity and the targeted pattern coverage (gain).

To undertake this issue, the proposed model involves the following contribution.

• **Contribution #3:** Developing a cost-gain tradeoff measure to identify cost and gain settings that deliver the best possible refinement outcome.

1.1.1.4   Scenario-Driven Situation Refinement

Although the process of situation refinement has a common goal in reaching a higher level of situational awareness, most CEP tuning methods follow different directions and approaches toward reaching that goal. The reason for such diversity is due to several factors, such as the type of domain knowledge, the existing system architecture, and the data mining technique used for the rule-learning task.

Although these factors have been extensively investigated in related literature, yet, the nature and characteristics of the real-world scenario targeted for refinement are rarely addressed.

Since the situation refinement process intended to target several types of real-world scenarios that vary in their business goals and requirements; hence, situation refinement process must adapt and align with such requirements.

Therefore, the proposed CEP-tuning model must adopt the business needs to represent the targeted situation accurately. This can be achieved by formalizing and incorporating particular scenario requirements as input parameters in the proposed refinement criteria.

To undertake this issue, the proposed model involves the following contribution:

• **Contribution #4:** Using Refinement Mode Factor to adjust cost evaluation criteria to the targeted refinement scenario.

1.1.2 Contribution Summary

Table 1.1 shows a summary of the proposed model's contributions identified in Section 1.1.1.

Table 1.1: **book** Contribution Summary

Contribution No	Discription	Key Feature
1	• Enriching CEP rules with the current state of knowledge to represent the user's preferred context.	*Context-Aware*
	• Using a Custom Support Threshold to govern the level of situation refinement.	*Evemt Streaming Compliant*
2	• Using a Relative Weight Factor to re-evaluate the mining outcome as the event stream evolves.	*Event Streaming Compliant*
3	• Using a cost-gain tradeoff measure to deliver the highest refinement outcome.	*Cost-Based*
4	• Using Refinement Mode Factor to adjust the cost evaluation criteria for the targeted refinement scenario.	*Scenario-Driven*

1.2   Problem Solving Approach

To fulfill this work's objectives, we use a multidisciplinary problem-solving approach,
as shown in Fig. 1.2. This approach involves narrowing the scope of work as each
milestone is reached.

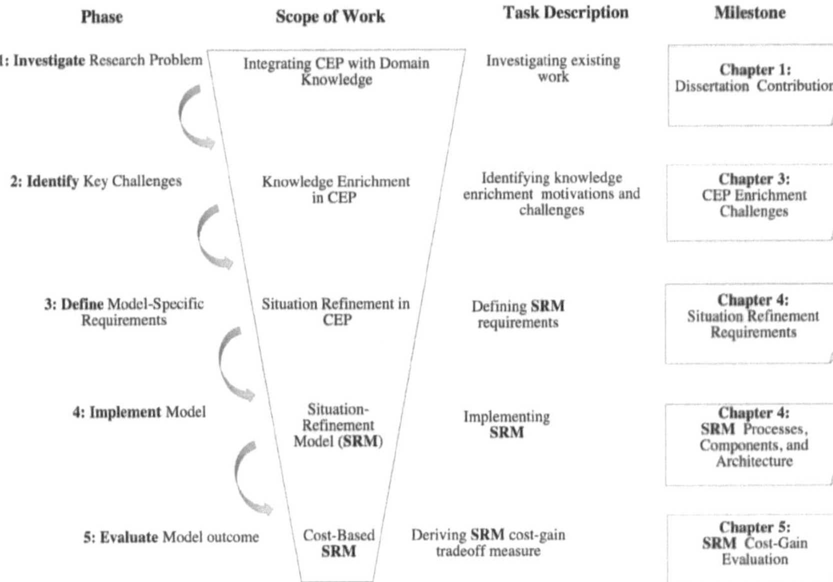

Figure 1.2: Problem Solving Approach

Chapter 2

Background

2.1 Complex Event Processing: An Overview

Since its advances in the last two decades, CEP technology become a core component in many application areas that require a higher level of real-time situational awareness such as Business Automation, Smart Environment, Item tracking, and Traffic management.

Complex event processing is a set of techniques and tools that combines data from event streams to identify meaningful patterns of events (such as opportunities or threats to the system) and responds to them as quickly as possible. In abstract terms, CEP is an approach that identifies data and application traffic as 'events', correlates these events to reveal predefined 'patterns', and reacts to them by generating 'actions' to systems, people and devices [29].

Complex event processing (CEP) provides an effective solution to process real-time event streams for many of today's business applications; as a result, complex event processing becomes a mainstream approach for processing streams of data [4]. To achieve this, CEP systems rely on predefined complex patterns templates to detect situations of interest from ongoing event streams(s). Complex pattern templates contain relational operators and variables that compose detection rules required to identify a particular situation as it occurs.

2.1.1 Event Related Terminology

For this work, the event-related terminologies used throughout this work complies with Complex Event Glossary Version 2.0 [30]:

- **Event**: An object that represents, encodes, or records an event, generally for computer processing.

- **Event attribute** (event property): A component of the structure of an event.

- **Simple event**: An event that is not viewed as summarizing, representing, or denoting a set of other events.

- **Complex event**: An event that summarizes, represents, or denotes a set of other events.

- **Complex event processing** (CEP): Computing that performs operations on complex events, including reading, creating, transforming, abstracting, or discarding them.

- **Event pattern**: A template containing event templates, relational operators, and variables.

- **Event stream**: A linearly ordered sequence of events.

2.1.2   Complex Event Processing: Main Tasks

Functionally, CEP technology collects a variety of information from incoming events as they occur. It provides an organization with the ability to define, manage, and predict complex patterns of events in near real-time. Therefore, CEP responses are not triggered by a single event. Instead, it is triggered by a complex composition of events happening at different times and within different contexts. As depicted in Fig. 2.1, complex event processing is performed in three stages as follows:

1. **Event capture**: the CEP engine captures the streams of incoming events.

2. **Event analysis**: this stage involves using event patterns to process and correlate event information to detect situations of interest.

3. **Response**: As soon as a complex event is detected, the CEP engine notifies the system or event consumer about a particular situation of interest.

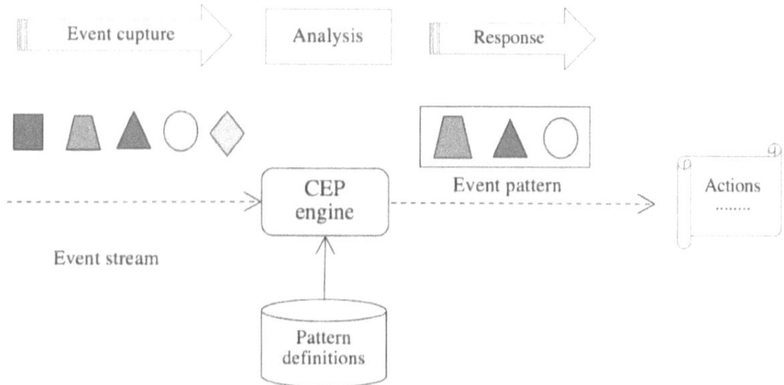

Figure 2.1: Complex event processing stages [1].

2.1.3 Complex Event Processing: Main Application Areas

Complex event processing has become a core technology to process a large flow of event data to detect situations of interest and respond to them as quickly as possible. Consequently, complex event processing has replaced traditional database systems in many mission-critical application areas such as:

- **Sensor Networks**: Detecting emerging situations (e.g., fire) within industrial facilities by monitoring and handling a large amount of stream data as they generated through sensory networks.

- **Algorithmic Trading**: A CEP engine is used to infer possible fraudulent transactions based on a predefined complex pattern represent clients' purchase behavior.

- **Business Activity Monitoring**: CEP technology can be used to define and analyze opportunities and risks within an organization proactively.

- **Traffic management**: Complex Event Processing can be used to detect vehicular occupancy along with the meaningful points of the frequent itineraries based on a density-cluster algorithm.

A primary motive for using CEP technology in such systems is to reach a higher level of situational awareness within the application domain by providing better insights about emerging complex situations. To achieve such a goal, it is essential to detect complex situations within the application context, i.e. , inferring complex events based on the correlation between event data and background knowledge [27].

From the Situational Awareness (SA) viewpoint, obtaining the state of knowledge by recognizing a particular *SOI* requires gathering sufficient related facts within a specified time and space [7–9]. For example, in smart grid applications, enriching event data with the background knowledge that represents energy producers, grid operators, or business customers can help decision-makers to improve their awareness about emerging situations [7, 31].

Technically, complex event enrichment is the process of fusing background knowledge with the event stream to facilitate the event processing engine to learn more about incoming events and their relationships with other environmental concepts [14, 32].

2.2 Situation and Situational Awareness in Literature

The earliest formal presentation for the notion of the situation was introduced by Barwise [33] to define system properties such as relations, events, and event situations as objects. According to Barwise, a situation corresponds to a limited number of objects of the real-world that humans perceive and reason about. Others developed formal semantics to represent situations by utilizing real-world objects, including von Mises [34] and Bunge [35].

Although these semantics follow different approaches to represent the situation, they mostly agreed that the situation must be realized based on the knowledge of the surrounding system objects, relationships among these objects, and the events that define them.

The advancement in real-time sensing technology and the use of data streams in in the late twentieth and early twenty-first centuries show the importance of expanding the notion of a situation to fulfill real-time streaming requirements, i.e., by maintaining the change in the state of system objects instantaneously and continuously.

Therefore, several studies in the Data Fusion community used the term Situational Awareness to address this issue. These studies were based mainly on the *Endsley* mental model [8]. Endsley defined situational awareness as the "perception of environmental elements and events with respect to time and space, the comprehension of their meaning, and the projection of their future status" [36]. As shown in Fig. 2.2, *Endsly* model involves three levels of presentation to achieve Situational Awareness: perception, comprehension, and projection [2].

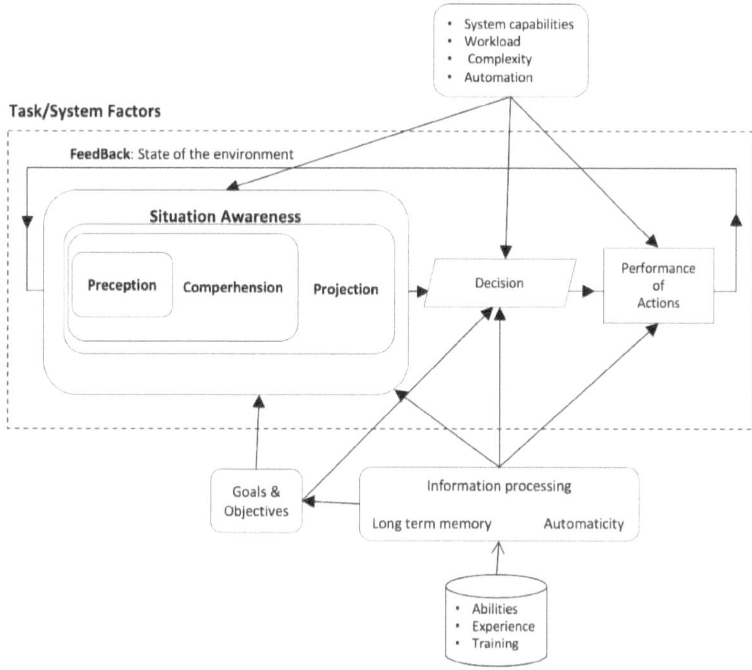

Figure 2.2: *Endsley's* Situation Awareness Model [2]

Among a large number of data fusion models proposed to outline the concept of situational awareness, the data fusion model proposed by the Joint Directors of Laboratories Data Fusion Group (JDL) [37] is the most widely accepted standard, which provides a detailed formalization of the situation and its sub-concepts. Fig 2.3 shows five levels of the JDL Data Fusion process.

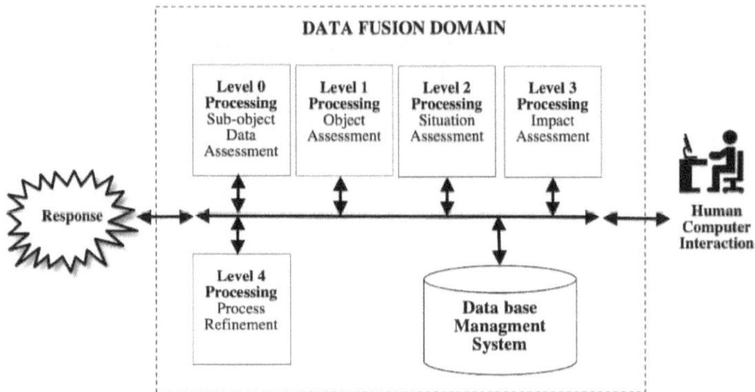

Figure 2.3: JDL Fusion Model [2]

2.3   Situational Awareness and CEP: Challenges

As the CEP technology becomes a core component in many application areas nowadays, several related studies have focused on the adaptation of the JDL data fusion
model into CEP [38–41]. Despite partial success in this direction, though, the JDL
model's applicability to the CEP paradigm faces two major challenges: First, the
process of enriching CEP detection rules with external domain knowledge requires
an efficient information-fusion mechanism capable of processing a vast amount of information in near real-time [10,27], and second, the information fusion process must
take into account the limited time available to process information from knowledge
sources that are heterogeneous by nature, such as ontologies, linked data, etc. [42–44].

Dealing with these challenges requires enriching CEP patterns with a large number of detection rules in near real-time. From the situational awareness perspective,
the increase in detection rules makes the targeted situation more complicated to comprehend and handle [20,26]; furthermore, it leads to higher levels of computational
complexity [27].

To provide an in-depth description of these challenges, the rationale behind complex event enrichment, related challenges, and the existing approaches to tackle these
challenges have been investigated in our work in [10]. Chapter 3 illustrates this work.

Chapter 3

Complex Event Processing Enrichment: Motivations and Challenges

3.1 Introduction

Using CEP technology to detect complex situations early within the application context requires the utilization of various types of external knowledge sources. Therefore, external knowledge in the application domain is needed to enrich event data and to identify CEP rules towards detecting critical situations in near real-time [27,45].

CEP enrichment is the process of fusing background knowledge with the event stream to facilitate the event processing engine to learn more about incoming events and their relationships to other related concepts [14,32].

The notion of event enrichment has been explicitly mentioned in several studies, along with the rise of complex event processing [3,46–49]. The significance of event enrichment presented as an essential feature that CEP based applications must possess to interpret received events by combining them with external knowledge sources.

Despite significant improvements in applying various knowledge discovery techniques to retrieve CEP domain knowledge, such a process still faces several limitations to achieve this goal. In particular, the compliance with event-streaming restriction concerning time [12,50], and the challenge of developing a unified retrieval mechanism capable of processing various types of information in the external knowledge sources [43,44].

In this Chapter, we address the motivations and challenges facing CEP enrichment. To do this, we present a motivation oriented classification of CEP enrichment techniques and approaches as exhibited in our work in [10].

3.2 Motivations Behind CEP Enrichment

Due to its loose-coupling nature, CEP technology is used for various scenarios and to benefit from shared knowledge sources such as ontologies and linked data. As these scenarios differ in their goals and needs; hence, it is essential to understand the needs and motivations for a particular scenario to implement the proper knowledge-enrichment approach.

The CEP enrichment process is influenced by several factors such as application area targeted for knowledge enrichment and knowledge source update and retrieval mechanisms [7,32,50]. In the following subsections, the motivations for complex event enrichment are presented in detail.

3.2.1 Real-time Situational Awareness

A primary motive for enriching CEP systems is to provide better insights about emerging complex situations by developing the relationship between event data flows and current business changes reside in background knowledge [7,45]. In other words, the aim is to achieve a higher level of situational awareness by inferring more expressive complex events based on the correlation between event data and background knowledge [27].

3.2.2 The Need to Update CEP Rules Over Time

Usually, CEP systems depend on domain experts to specify user-defined rules for complex patterns [23,51]. These rules must be modified by human experts to identify and express correlations between the various events and situations of interest. Providing all the required information to perform rule updates manually is a hard task, especially with the need to perform it continuously with low latency [11,21].

Therefore, several critical-mission scenarios, such as Credit Card Fraud or Trends Discovery in the Stock Market tend to tackle such bottleneck by developing customized rule-tuning methods to match their application needs [11,13,52].

3.2.3   Decoupling Nature of CEP-based Systems

As an event-based paradigm, the CEP-based systems support the concept of decoupling among various interacting components involved, such as event producers, event consumers, and shared knowledge sources [32].

In general, shared knowledge sources depend on other systems and means beyond the scope of CEP engine to update their state (e.g., ontologies and linked data). Therefore, it is essential to have full synchronization between knowledge sources and CEP detection rules to detect emerging situations based on the current state of domain knowledge.

3.2.4   Incomplete Primitive Event Information

A large number of CEP-based applications nowadays tend to use event sources with limited functionality (e.g., *RFID* readers) to maximize throughput and minimize latency [32,53,54].

These event sources reserve a limited space in memory to hold only basic event data to cope with high-speed stream processing and to avoid data-queuing. Therefore, raw event data must be enriched with complementary knowledge required to recognize and discover situations of interest.

3.2.5   The Transition from Reactive to Proactive Complex Event Processing

Proactive event processing is the ability to discover or eliminate undesired future events (e.g., false alarms) or to identify future opportunities [53]. Proactive event processing is based on the idea of automated generation of CEP detection rules to predict future event occurrences [11,55].

Proactive event processing can be achieved by applying machine learning and predictive analytics techniques to learn and anticipate future event occurrences based on historical event data and up-to-date knowledge acquired through CEP enrichment.

In transportation systems, for example, possible future traffic congestions can be anticipated and possibly prevented by enriching CEP detection rules with resultant predictions learned from past event occurrences.

3.2.6 Semantic Enrichment

One major motive to enrich CEP detection rules is to detect and process raw events semantically to create more expressive complex events [12, 13]. Semantic enrichment is the process of expressing CEP patterns based on business concepts, relationships, and rules about the particular application domain, which can be inferred from shared knowledge sources (e.g., ontologies, linked-data) [56, 57].

Two advantages can be gained from semantic enrichment: First, complex patterns and reaction rules can be defined and expressed precisely using business terms and vocabulary; and second, re-using domain knowledge (update once and use many).i.e., whenever the business rules are updated in the knowledge base, the update is reflected instantly in CEP detection rules [58, 59].

Nowadays, several application areas benefit from semantic enrichment, such as SmartGrid, Health Care, Logistics, and the Stock Market [57, 60–62].

3.2.7 Context Aware Complex Event Processing

A key factor in improving the quality of the decision-making in CEP systems is to react to the detected situations based on a particular application context [63–65]. In broad terms, context is a relevant knowledge property of the system that can be used to characterize the situation of a particular object or concept in the surrounding environment, such as a person, place, and time [66, 67]. In weather monitoring, for example, the knowledge property *'user-location'* can be used as a contextual property to describe the current temperature in *Fahrenheit* if the user is located in the US.

In the event streaming environment, context awareness can be defined as the ability of a particular system to analyze and describe the detected situations based on the user-preferred context, among other available contexts, to provide the appropriate reaction in near real-time [68, 69].

The benefits gained from enriching CEP systems with the knowledge properties that represent a particular context is a strong motive to develop context-aware solutions in application areas such as Smart City [70], Internet of Things (IoT) [54], Energy Management [7], and Vehicular Control [63]. In Call Management, for example, context can be used to redirect received calls to the appropriate agent based on spatial context (location) or a social context (language, education, etc.).

3.3   CEP Enrichment: Motivations vs. Approaches

The motivations presented in Section 3.2 show a strong foundation in applying complex event enrichment. However, it is rare to find an existing CEP enrichment approach driven by a single motivation.

In fact, most of the current CEP enrichment solutions are driven by multiple motivations. This can be seen in many existing solutions, such as *Vehicular Occupancy Detection* [63] with motivations as discussed in (3.2.1, 3.2.4, 3.2.5, 3.2.7), *Road Traffic Management* [71] driven by motivations (3.2.1,3.2.6,3.2.7), and the *IT-telecom convergence* [58] motivated by 3.2.2,3.2.6,3.2.7. Other examples can be found in [7, 13, 25, 72].

In our work, a total of 20 existing solutions in 4 application areas have been investigated to identify possible correlations between the application area and the motivations.

Table 3.1 shows the various motivations behind CEP enrichment development in these use cases. Fig. 3.1 shows the percent distribution of CEP enrichment motivations for the use cases under investigation. From Table 3.1, the following indications are observed:

1. Situation Awareness, Semantic Enrichment, and Context Awareness are the leading motives for CEP enrichment, as they appear in almost all the use cases.

2. There is a strong association between Semantic Enrichment and Context Awareness as both exist jointly in several use cases. This correlation is due to the use of both Context-Awareness and Semantic technologies to enrich event data.

3. Situation Awareness is the main drive for enrichment in two particular application areas: Smart Environment and Social Networks. The existence of Situation Awareness as a motive can be noticed in almost all use cases in these application areas.

Table 3.1: CEP Enrichment: Application Area verses Motivations

No	Application area	Use case	Motivation							Ref
			Situation Awareness	Updated CEP rules	Decoupling nature	Event Enrichment	Proactive CEP	Semantic Enrichment	Context Awareness	
1	Business Automation	Process management					✓	✓	✓	[73]
2		Container inspection		✓				✓		[13]
3		Parcel delivery		✓				✓		[74]
4		Transport & logistics					✓			[75]
5		Hosptial care						✓	✓	[25]
6	Smart Environment	Energy management	✓					✓	✓	[7]
7		Smart grid analytics		✓	✓	✓		✓		[42]
8		Smart cities	✓	✓				✓	✓	[70]
9		Smart Homes	✓		✓		✓			[76]
10		Renewable energy	✓				✓	✓		[60]
11	Social Networks	Disease monitoring	✓					✓	✓	[77]
12		Emergency warnings	✓	✓				✓	✓	[78]
13		Community detection	✓						✓	[79]
14		News feeds processing	✓					✓	✓	[80]
15	Traffic, Vehicle control	Vehicular control	✓			✓	✓		✓	[63]
16		Traffic management						✓	✓	[71]
17		Vehicular routing						✓	✓	[43]
18		Traffic simulation				✓	✓			[81]
19		Traffic analysis		✓		✓				[11]
20		Traffic management						✓	✓	[82]

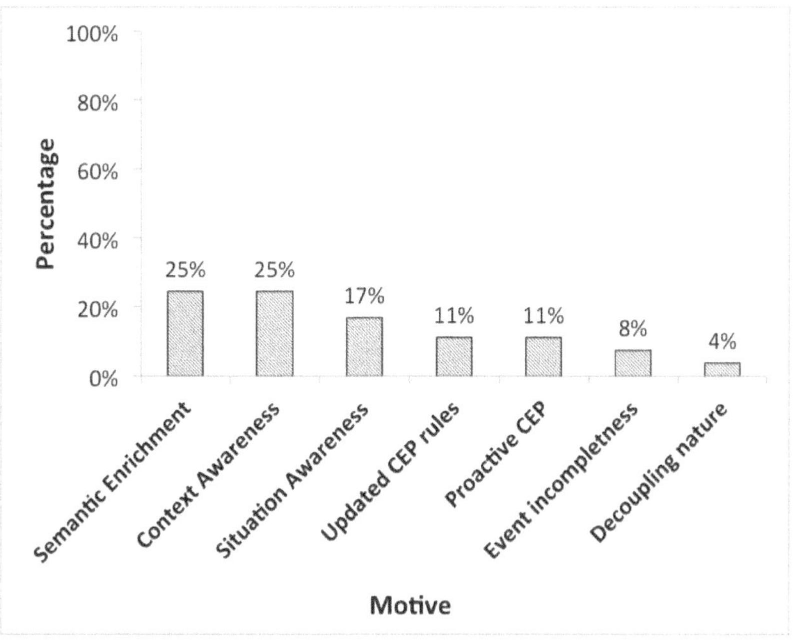

Figure 3.1: Distribution of CEP enrichment motivations

3.4   CEP Enrichment Challenges

As mentioned earlier, the process of enriching complex events with external knowledge must be accomplished in compliance with event streaming requirements. Therefore, two restrictions must be taken into account: First, due to the unbounded nature of event streams, CEP detection rules must be updated continuously [51, 83], and second, since event streaming requirements imply a low latency and high throughput, the knowledge fusion process must be accomplished in near real-time regardless of the event stream's speed and size [27, 84].

Several challenges must be undertaken to tackle these restrictions. In particular, human intervention, Big Data challenge, real-time stream reasoning, and knowledge representation challenge [23, 32, 85].

To associate these challenges to the use cases under investigation, Table 3.2, and Fig. 3.2 shows the distribution of CEP enrichment challenges.

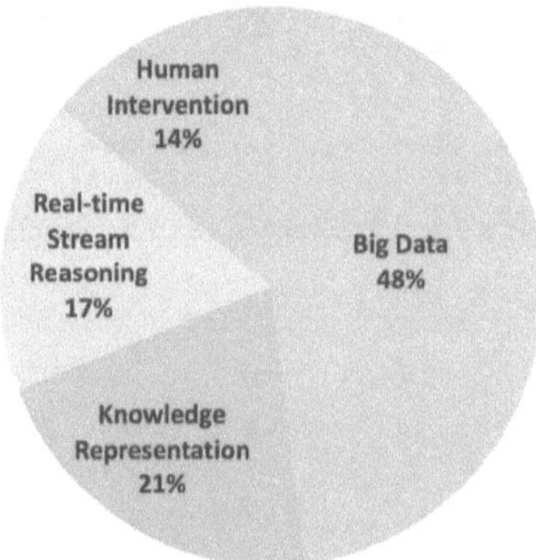

Figure 3.2: Distribution of CEP Enrichment challenges

3.4.1   Human Intervention Factor

One of the main issues facing CEP systems is the need to update CEP rules continuously by human experts. Performing such a task within the event streaming environment is challenging especially in mission-critical systems where human experts are required to handle an enormous amount of domain knowledge in a near

Table 3.2: CEP enrichment: Application Areas verses Challenges

No	Application area	Use case	Challenge			
			Human Intervention	Big Data	Real-time Stream Reasoning	Knowledge Representation
1	Business Automation	Process management		✓		
2		Container Inspection	✓		✓	
3		Parcel Express Delivery		✓		
4		Transport & Logistics		✓		
5		Hosptial Care		✓	✓	✓
6	Smart Environment	Energy Management		✓		
7		Smart Grid Analytics		✓		
8		Smart Cities	✓		✓	
9		Smart Homes		✓		✓
10		Renewable Energy		✓		✓
11	Social Networks	Disease Monitoring		✓		✓
12		Emergency Warnings	✓	✓		
13		Community Detection		✓		
14		News Feeds Processing		✓		
15	Traffic & Vehicle Control	Vehicular Control				✓
16		Traffic Management		✓		✓
17		Vehicular Routing			✓	
18		Road traffic Simulation				
19		Vehicular Traffic Analysis	✓	✓		
20		Traffic Management			✓	

real-time and a continuous fashion.

In this direction, several approaches proposed to automate the process of learning and updating CEP rules. For example:

- Automatic learning of predictive rules [21].
- Automated rule generation from historical events [11].
- Sequence clustering-based rule generation [86].
- The rule tuning model using machine learning techniques [23].

Although the approaches above have gained some degree of success in tackling this issue concerning time, yet, the need to fuse updated rules with domain knowledge given the time constraint is still a limiting factor due to the heterogeneity of such knowledge.

3.4.2 Big Data Challenge

As most external knowledge sources hold Big Data characteristics (i.e., volume, velocity, and variety), therefore, CEP rules must be updated in synchronization with any change occur in the enrichment sources [42,87]. The significance of adopting such a policy can be seen in many critical-mission applications such as Threat Detection, Competitive Intelligence, and Disaster Management [71,77,78].

In these applications, immediate and accurate responses require keeping CEP rules up-to-date with the recent changes at the domain knowledge. The challenge here is that the knowledge update occurs in near real-time, from a variety of domains, and with a large volume of data [88]. Several CEP enrichment approaches have been proposed to cope with the Big Data challenge, for example,

- *Apache Storm*, an open-source data processing paradigm to process real-time Big Data streams [89].
- *SCEPter*, a framework that addresses Big Data challenges by processing CEP queries across real-time event streams [42].
- A CEP based publish/subscribe paradigm is introduced for Big Data analytics [90].
- *H2O*, Complex event processing framework utilizes Big Data sources to enrich event data semantically [88].
- Other solutions in this direction have applied in several use cases [25,71,78].

3.4.3   Real-Time Stream Reasoning

Real-time stream reasoning is the process of deriving and extracting the knowledge required to make decisions near real-time by using both heterogeneous data sources and event streams [45,91]. The reasoning process is a core part of any CEP system that utilizes ontology and RDF knowledge sources to decode and transform event data into a useful knowledge semantically [43,92]. Since event streams are unbounded, the reasoning process has to be done incrementally as new information becomes available.

Two significant bottlenecks face CEP systems to accomplish stream reasoning in near real-time. First, a large number of instances in the ontological knowledge sources must be processed in a small time-window [19,93]. Second, the increased computational complexity as a result of performing reasoning tasks incrementally [18].

Several existing methods gained different degrees of success in tackling the bottlenecks above, for example:

- *EP-SPARQL*, a semantic query language for complex events and stream reasoning [91].
- *C-SPARQL*, a semantic language for continuous queries over streams of *RDF* data [94].
- A distributed reasoning method to support distributed reasoning on expressive background knowledge [95].
- *OECEP*, a reasoning approach that combines complex event processing and domain knowledge from ontologies [13].

3.4.4   Knowledge Representation Challenge

As shown in Section 3.2.3, enrichment knowledge sources are loosely-coupled with CEP systems to cope with scalability and integration requirements. These knowledge sources differ in their semantics, and the mechanisms used to describe, update, and retrieve knowledge required for event enrichment [32].

The main issue with this regard is the Knowledge Representation challenge.i.e., the need to represent and assess discovered situations using domain-specific vocabulary. Achieving such a goal requires processing a large number of CEP rules in a small time-window to guarantee the fusion between low-level data (raw events) and high-level, semantically-rich knowledge sources [7,61,96].

Bridging the gap between event streaming restrictions concerning time and the information fusion requirements concerning knowledge representation has been under focus in both Data Fusion and CEP community. The main achievement in this direction is the introduction of the *JDL* Fusion model (discussed in Section 2.2), which provides a multi-layered view that combines both low-level information fusion and high-level Knowledge Representation.

Nevertheless, the *JDL* model's applicability in the event streaming environment still faces two significant bottlenecks: the small time-window available to fulfill the whole Information Fusion process, and the amount of computation required to process an increased number of CEP rules as the event stream evolves. Despite that, several solutions have gained a partial success in tackling Knowledge Representation challenge, for example,

- A dynamic enrichment paradigm based on semantic similarity for large-scale and open web sensor networks as introduced in [32].
- *Approximate Semantic Matching*, a model that uses the Approximate Matching technique to reduce the Knowledge Representation computational cost [61].
- Other attempts tackle the Knowledge Representation challenge partially [25,60,71, 97].

3.5   Chapter Summary

This Chapter presents the motivations and challenges facing CEP enrichment, as illustrated in our work in [10] . In this work, 20 related use cases have been examined to identify the possible correlation between CEP enrichment challenges and the motivations behind these solutions. Based on the results, the following indications are observed:

1. Improving situational awareness about emerging situations implies increasing CEP pattern rules to reach the proper level of Knowledge Representation.

2. The need to enrich CEP systems continuously implies tackling the issue of the increased number of CEP rules as the event stream evolves.

3. The increased number of CEP rules makes *SOI* presentation more challenging to comprehend and handle by the user; furthermore, it increases the consumption of available computational resources required to store and process these rules.

By taking into account all the above, presenting *SOI* at a proper level of Knowledge Representation implies a tradeoff between the number of CEP rules required to express *SOI* and the desired level of knowledge representation. Such a tradeoff is essential to avoid the issue of situation complexity as a result of an increased number of rules.

In Chapter 4, we present **SRM**, a situation refinement approach that considers issues mentioned above while respecting real-time event streaming constraints.

Chapter 4

A Situation Refinement Model for Complex Event Processing

In this Chapter, we present a Situation Refinement Model (**SRM**) to fine tune CEP-rules based on the tradeoff concept discussed in Chapter 3. The proposed approach aligns with the **book** contributions #1 and #2 (Section 1.1). The official manuscript of the **SRM** model can be found in [28].

4.1   Introduction

Identifying a particular *SOI* that implies an opportunity or threat, requires gathering sufficient facts in near real-time from surrounding domain knowledge to support the existence of that situation [10–12]. Therefore, it is essential to have an efficient data stream mining algorithm to infer, analyze, and deliver the required knowledge within the specified time limit.

The main issue related to identifying situations in near real-time is the continuous need to learn and update CEP detection rules to adapt to the emerging situation in near real-time [11, 13, 14]. Hence, any data stream mining mechanism used for such a purpose must have a high degree of adaptation to the changes in situations of interest. Specifically, the process of rule mining must comply with event streaming constraints, i.e., the input events must be acquired in near real-time and with a single-scan mode to cope with the continuous arrival of data elements and to avoid data queuing [14–17].

Several research efforts pointed out the challenge of applying data mining techniques with limited time available to learn CEP detection rules from the real-time event streams [18, 19]. Existing rule learning approaches either focus on mining CEP rules from historical instances of events [20 23] or surrounding domain knowledge [13,14]. Other attempts focused on combining domain knowledge with historical events in the rule learning process, such as the *SCEPter* framework proposed by Zhou et al. [24].

Despite the partial success in automating the CEP rule generation process either by fine tuning existing CEP rules [23, 25] or by generating entirely new pattern rules [20, 21], dealing with the increased number of rules is still a challenging task. Precisely, the increased number of CEP detection rules leads to more complex detection patterns, which makes the targeted situation more complicated to comprehend and handle [12,20,26]; furthermore, such an increase also brings higher levels of computational complexity [27].

On the other hand, limiting the number of CEP rules might not provide enough information that reflects the desired level of situation refinement. To sum up, tackling these challenges implies achieving an acceptable tradeoff between situation complexity and the number of CEP rules needed to identify that situation. This leads us to the primary question related to our work:

How to gain the desired level of situation refinement while maintaining the minimum number of CEP rules?

Motivated by the issues stated here, we present **SRM**, a Weighted Frequent Item Mining (*WFIM*) approach to identify and select the minimal set of CEP detection rules required to refine a particular *SOI*. The proposed model incorporates the following characteristics:

1. Inferring and learning pattern detection rules based on their frequency of occurrence in the event stream and their state-values in domain knowledge.

2. Using user-defined custom support threshold to govern the level of situation refinement while maintaining the minimum number of CEP rules.

3. Using a Relative Weight factor to accurately update CEP rules statistics as the event stream evolves.

One of CEP technology's capabilities is its decoupling nature that allows processing, analyzing, and extracting complex patterns in near real-time with a distributed architecture. Such capability exists due to two main reasons: First, the CEP ability to correlate event streams from different data sources using distributed message-based systems, and second, CEP engines can be distributed across a cluster of machines. For this work, the **SRM** interacts with the CEP engine in an ad-hoc manner to refine complex pattern specifications.

CEP technology that supports the Event-Condition-Action paradigm (ECA). It supports event handling, event detection, and the actions to be performed due to event detection. To fulfill these tasks, CEP technology uses two types of rules: Event detection and Action rules. The event detection rule is the conditional expression in the complex pattern query that specifies the *SOI* to be detected. The action rules are related to the decision-making process in CEP, which involves the task of defining and executing the actions through action rules. The **SRM** scope of work only involves refining event detection rules to detect a particular *SOI* and to deliver feedback to the user as an alert or notification in near real-time; therefore, the task of decision analysis and decision making is out of the **SRM** scope of work.

4.2 Situation Refinement: Problem Nature and the Proposed Measures

Several Frequent Item Mining models (*FIM*) are proposed to tackle the issue of the increased number of mined items in real-time event streams [84, 98–100]. Despite their success in applying different measures and techniques to tackle such a challenge in various real-life event streaming scenarios, these approaches have their limitations in CEP-based situation refinement scenarios. The following subsections illustrate the nature of the Situation Refinement problem and its significance along with the unique restrictions that need to be applied during the refinement process.

4.2.1 Pattern Coverage Versus Pattern Complexity

In general, Frequent Item Mining (*FIM*) models deal with the increased number of items on the unbounded data streams either by selecting items according to a particular frequency threshold or a maximal item threshold [101–104]. Using either or both thresholds during the mining process produces a fixed outcome concerning the resultant set of items and the overall coverage, therefore, whenever a user wishes to change the threshold value(s), another scan to the whole input event stream is required.

In a car sales scenario, for example, if the existing situation S indicates an increase in sports car sales during a particular period, then, a user may wish to have different representations (tradeoffs) of the targeted SOI such as:

SOI: "40% of S-car sales are of *German* origin."

SOI: "70% of S-car sales are either *German* or *French* origin."

SOI: "85% of S-car sales are either *German*, *French* or *Japanese* origin."

Achieving such tradeoffs from evolving event stream requires storing the whole event data to analyze and extract the required items in near real-time; however, this is impractical since the event stream is continuously evolving. Further, the leverage of re-visiting past event data is not an option due to the real-time streaming constraints and big data characteristics [84, 99, 105, 106].

To avoid re-visiting past event data, the proposed refinement model uses the Gradual Inclusion Approach to gather, analyze, and accumulate the current rule statistics and added to the past rule summary as the event stream evolves.

4.2.2 Governing the Outcome of Situation Refinement Process

Although representing SOI with different tradeoffs meets the needs of many Situation Refinement scenarios, other scenarios may require the opposite approach, i.e., managing the refinement process based on a specific outcome. Precisely, a manageable number of complex events must be maintained at any given time regardless of the size of the event stream.

In customs inspection, for example, it is a challenging task to react to all complex event notifications that indicate a request to perform the container inspection procedure, due to the limited time and resources available to perform such a task. Therefore, system owners aim at maintaining a manageable number of containers at any given time to react appropriately to a given situation in near real-time.

The challenge here is how to provide the user with the desired tradeoff that meets the required outcome. In other words, maintaining a fixed outcome as the event stream evolves.

To achieve a fixed outcome, the **SRM** model uses a Minimum Support Threshold (MST) as an input parameter to govern the level of Situation refinement. To ensure that the selected CEP rules are minimal, the model gradually accumulates CEP rules based on their relative weights in descending order until the refined pattern coverage reaches the *MST* threshold.

4.2.3 Maintaining CEP Rule Statistics for a Dynamically Changing Environment

Due to the dynamically changing nature of the event streams and the change in seasonal trends over time, the process of mining CEP rules must reflect the imbalanced distribution of the event stream as the stream evolves. To accurately reflect the change in event stream trends in the targeted *SOI*, the average rule summary in dense periods must gain more weight than the less dense periods. Therefore, the average rule summary for the current time-window must be calculated individually in near real-time and then accumulated to past rule summary.

Since each time-window represents a different weight, hence, accumulating the current window's statistics to the past summary requires keeping all past time-windows' statistics to re-calculate the overall rule summary at the end of each time-window. Nevertheless, such leverage is not an option due to the limited memory, the amount of computations required in near real-time, and the continuous evolvement of the event stream.

To the best of our knowledge, none of the existing Frequent Item Mining (FIM) time models fulfills the above mentioned requirements, such as the *sliding-window* model that considers the most recent window only [107–109]. In the *land-mark* model [110,111], equal importance is assigned to all mined items in the stream; still, it does not consider the imbalanced distribution of the event stream. Also, in the *time-fading* model, the entire event stream is taken into account; though, the most recent part of the stream flow has more weight than the older one [112,113].

To accurately reflects the change in event stream over time, the **SRM** model uses a Relative Weight factor λ to re-distribute and update the overall rule summary as the event stream evolves.

4.3 Related works

This Section provides a detailed description of related existing work. The various approaches used to learn CEP detection rules by incorporating domain knowledge with historical instances of events illustrated in 4.3.1. The approaches and techniques associated with rule mining over data streams presented in 4.3.2.

4.3.1 Integrating CEP with Domain Knowledge Toward Situation Discovery

Incorporating domain knowledge captured from external sources to enrich CEP systems has been well studied over the past decade [12, 27, 50, 114].

Several related studies and surveys reveal the importance of CEP enrichment in gaining a higher level of real-time Situational Awareness [10, 30, 50, 115, 116]. Other CEP enrichment approaches focus on fusing complementary knowledge with event data upon the arrival of raw event data [7, 117] or upon complex event detection [118]. Hassan et al. [7] addressed the challenges associated with achieving a higher level of Situational Awareness in CEP systems by proposing a CEP enrichment model that combines sensor network information flow with domain knowledge from linked data cloud.

Other research efforts focus on developing enrichment approaches based on the semantics of events, mainly by using ontological knowledge to enrich CEP rules. Teymourian et al. [27, 119] proposed Event Query Pre-Processing (*EQPP*) to semantically re-write pattern queries using available domain knowledge.

Turchin et al. [23] proposed a CEP auto-tuning model to update rule specifications over time. The authors used *Discrete Kalman Filters* to specify and tune rule parameter values. Mutschler et al. [22] introduced a method for learning event detection rules using hidden Markov models (HMM). Similarly, other methods proposed to learn CEP rules from event history automatically, such as *iCEP* [11], *autoCEP* [21], and *SCARG* [120].

Several studies tackled the issue of automatic rule generation by combining Predictive Analytics (PA) techniques with CEP to predict proactively future situations [10], for example, the *CEP-PA* framework [121], *CEODE* [122] and *AMWR* [123, 124].

The synergy between smart environment, sensing technology, and visualization tools have empowered some CEP based applications with the ability to model, observe, and visualize interesting situations [125–128]. The most notable achievement is the *EventShope* [125, 129], a generic framework for situation modeling with a particular focus on integrating spatial-temporal aspects within heterogeneous event sources. Likewise, *StreamLoader* [128, 130], a web application with an interactive interface to visualize *SOI's* during the discovery period. Also, *MEdit4CEP* [1, 131], a domain-specific modeling language for CEP applications to handle and graphically configure CEP operators.

In summary, all mentioned enrichment approaches have unique strengths concerning fine tuning CEP rules to discover and present *SOI*; however, none of these methods fully comply with situation refinement requirements, as exhibited in Section 4.2.

Other notable weaknesses can also be observed in the presented approaches. For example, *SCEP* [27, 119] requires a partial user intervention during the situation discovery process. Also, *MEdit4CEP* [131] needs multiple scans to refine the situation conditions and parameters. Other methods, such as *SCARG* [120] and pattern mining approach [132], deal only with sequential patterns in the event stream while ignoring other types of pattern constructs such as conjunction and disjunction operators.

Nevertheless, it is worthwhile to mention that *iCEP* [11], *autoCEP* [21], and the CEP enrichment approach of [7] have successfully presented a CEP enrichment model using a reduced number of CEP rules and highlight the impact of rule complexity on the overall system performance. However, these methods do not address the uniqueness of Situation Refinement scenarios and the need to have multi-tradeoffs in the targeted situation as exhibited in Subsection 4.2.1.

The next Subsection highlights various mining methods and approaches used for rule discovery over data streams that consider the challenges and constraints associated with the real-time streaming environment.

4.3.2   Rule Mining Over Data Streams

Rule mining over data streams is a traditional problem for the data mining community due to its significance in several application areas such as retail market and stock market analysis [99, 108, 133–135].

Using traditional machine learning or neural network techniques to refine complex pattern detection rules faces two main obstacles. First, machine learning requires multiple data training cycles and testing to deliver the final rule set. In contrast, rule refinement in the event streaming environment must be in a single scan due to time and memory storage restrictions. Second, neural network algorithms involve using multiple and progressive network layers to extract higher-level features from raw input. However, rule learning in an event streaming environment can not leverage such an advantage due to time restriction and the continuous changes in system behavior as the event stream evolves.

As stated by Farzanyar et al. [84] and others [99, 105, 106, 136, 137], the following requirements must be satisfied by any data stream mining algorithm:

(1) A single scan to cope with the continuous arrival of data elements in the stream.

(2) A constant range of memory must be allocated to store mined elements regardless of the data stream size.

(3) Data elements have to be processed upon arrival to avoid data queuing.

(4) Datastream analysis results must be available as soon as possible to provide an adequate response in near real-time.

In general, Frequent Item Mining models use the frequency threshold or the maximal number threshold to limit the number of mined items on the stream. Additionally, *FIM* models maintain in-memory counters to keep occurrences of the selected items as the data stream evolves, such as *CP-tree* [138], *lambda-HCount* [139], and *PST-Miner* [101].

Nevertheless, these methods have limited ability in handling the increased number of items in the unbounded data streams. Such weakness exists because the number of items increases on a linear scale as the steam evolves, whereas the allocated memory has a limited size to cope with that increase [102, 103, 140].

Several *FIM* methods have proposed mining only the top-ranked elements (*top-k*) in the event stream, using the max-frequency measure to mine only the *top-k* items with the highest maximum frequency [141–143]. These approaches either focus on maintaining item frequency over the whole stream [144], or on recent stream activities [145–147]. Likewise, several Weighted Frequent Item Mining (*WFIM*) methods have focused on limiting the number of frequent items by using weight conditions to reflect a particular item's importance in comparison with other items [105, 148–150].

Other methods such as *MEIC* [151], *PrePost* [152], and *WEPS* algorithm [153], use the erasable pattern mining (*EPs*) technique to reduce the number of frequent patterns based on a user-specified threshold, or mining frequent patterns based on transaction value, such as the *HUPMS* algorithm [154] and *IWFPT* [155].

Existing Frequent Item Mining methods use different time models to maintain the mined item statistics during the discovery period, including the *land-mark* model, the *time-fading* model, and the *sliding window*. In the *land-mark* model [110, 111, 156], equal importance assigned to all mined items in the stream. In the *time-fading* model, the entire event stream is taken into account to compute each data item's frequency, but the most recent part of the streamflow contributes more than the older ones [104,112,113]. In *sliding window*, only data in the most recent window are taken into consideration [98,107–109].

Overall, existing *FIM* models either use frequency, maximal number, or the *top-k* threshold as an input parameter to limit the number of items during the mining process. However, the refinement requirements exhibited in Section 4.2.2 imply using the desired level of Situation Refinement as an input parameter in the mining process, i.e., maintaining a flexible number of items to represent the closest desired pattern coverage (outcome). Furthermore, the refinement requirements exhibited in Section 4.2.3 imply assigning equal weight to all mined items in the stream concerning the time of occurrence. Although the *land-mark* model satisfies such condition, yet, a weight factor must be used to signify the density of the various time frames during the discovery period, i.e., the total number of event occurrences in a particular time-window.

Based on the issues identified in subsection 4.3.1 and 4.3.2, the next Section presents the formal problem definition of Situation Refinement, and the proposed measures to rectify these issues.

4.4 Problem Definition

This Section presents the concepts and formal definitions of this work, along with the various factors required to refine CEP rules to identify a particular *SOI*.

4.4.1 Situation Refinement Concepts

Definition 4.1: *Situation (S).* Situation is referred to something happening in the real world, represented by the aggregation of multiple events from event stream E, occurred within a particular time period $T_{(t_1,t_2)}$, and detected by complex pattern P.

$$S = f\left(E, P, T_{(t_1,t_2)}\right),$$
where t_1 is the start time-stamp,

$\quad\quad t_2$ is the end time-stamp.

Definition 4.2: *Situation of Interest (SOI).* SOI is a refined observation of an existing situation S, supported by sufficient facts acquired from a particular domain knowledge D.

Refining the existing situation S to match the new *SOI* observation implies two constraints: First, maintaining the original situation characteristics by keeping the original pattern logical constructs, and second, adding new detection rules to the existing CEP pattern template to detect only event instances that match the targeted *SOI* observation.

$$\because SOI = f\left(S, D\right),$$

$$SOI \subset S$$

$$\therefore SOI = f\left(E, P, D, T_{(t_1,t_2)}\right)$$

Definition 4.3: *Rule.* A rule is defined as a conditional expression that specifies a single criterion to filter a particular set of complex events detected by the complex pattern P.

In the following example, pattern P uses the rule expression (Location=*'San Diego'*) to detect purchase events that occur only in *San Diego's* retail site.

P : **Purchase**(SalesList, Amount, Location)

$\quad\quad$ **WHERE** (Location=*'San Diego'*)

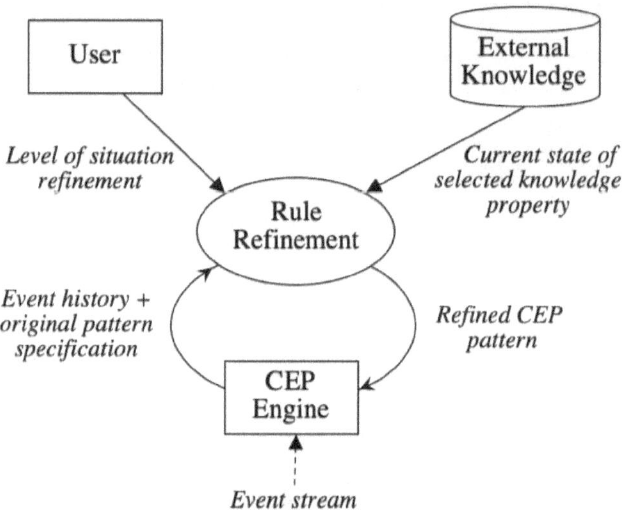

Figure 4.1: Rule Refinement Process

CEP pattern can combine more than one rule to refine its detection conditions to match specific criteria; this can be done using the fundamental logical operators (**AND, OR, NOT**), for example:

P :**Purchase**(SalesList, Amount, Location)
 WHERE (Location=*'San Diego'*
 AND Amount>500)

Definition 4.4: *Situation Refinement.* The process of learning and refining CEP pattern rules by incorporating domain knowledge and event history into complex pattern specifications to identify a particular *SOI*.

Fig. 4.1 shows the process of refining CEP rules to detect a particular *SOI*. Three inputs are required in this process: the current state of selected knowledge property, the desired level of Situation Refinement, and the event history that matches the original pattern P. The following scenario illustrates this process:

Scenario 1: A retail business usually sends a promotional offer to its customers who purchase more than 500 dollars. The logical expression of complex pattern P that detects eligible customers is as follows:

P: **Purchase**(SalesList, Customer, Amount)
    WHERE (Amount>500)

The business owners decide to reduce the number of eligible customers to 70% of its original size. To accomplish this, the business owners want to use a customer's loyalty factor (e.g., *Silver, Gold, and Platinum*) as a new selection criterion, i.e., by giving priority to the loyalty class whose customers have the most transactions first and then the subsequent ones until the desired rate is met.

Suppose that for a given period of T, there were 200 eligible transactions (with amount $500 or higher) detected by complex pattern P. Table 4.1 shows the distribution of these transactions based on the customer's loyalty factor.

Table 4.1:   Complex event distribution
(In descending order by # of transactions)

Loyalty	# of transactions where amount $\geq$ $500	Percentage %
'Platinum'	142	71.0 %
'Silver'	33	16.5 %
'Gold'	25	12.5 %
Total	**200**	**100 %**

To downsize the original number of transactions by 70%, a new detection rules must be added to reduce the original set of transactions to the closest value that matches the desired percentage. So, by adding the detection rule (LoyaltyState = *'Platinum'*), for example, pattern P detects only complex events that match the new detection rule. Consequently, the total number of detected complex events is reduced to 71% (Platinum customers percentage). Accordingly, the refined pattern P_{new} can be expressed as follows:

P_{new}: **Purchase**(SalesList, Customer, Amount)
    WHERE (Amount>500)
    AND (LoyaltyState = *'Platinum'*)

If the system user wishes to extend the detection criterion to match two customer loyalty classes (e.g., *Platinum and Silver*), both detection rules need to be unified in one logical expression.

Let P be the original pattern definition,

> P: **Purchase**(SalesList, Customer, Amount)
> **WHERE** (Amount>500);

Let R_1 be the logical expression that represents 1st rule (LoyaltyState = '*Platinum*').
Let R_2 be the logical expression that represents 2nd rule (LoyaltyState = '*Silver*').

By applying the 1st rule to the original pattern P, the new pattern P_{new} is as follows:

> P_{new}: $(P$ **AND** $R_1)$

Similarly, by applying the 2nd rule only,

> P_{new}: $(P$ **AND** $R_2)$

By unifying both expressions,

> P_{new}: $(P$ **AND** $R_1)$ **OR** $(P$ **AND** $R_2)$

> P_{new}: P **AND** $(R_1$ **OR** $R_2)$

Likewise, by unifying multiple rules,

$$P_{new} : P \textbf{ AND } (R_1 \textbf{ OR } R_2 .. \textbf{OR } R_r) \tag{4.1}$$

where r is the number of state-values required to refine pattern P.

According to definition 4.3, a rule R is composed of two components: the knowledge attribute k and the *value* that represents its state during complex event detection.

Definition 4.5: *State-value.* The value of a single attribute resides in the domain knowledge, which denotes the current state of a particular system object upon complex event detection.

As stated in the primary research question in Section 4.1, the challenge is how to gain the desired level of situation refinement while maintaining the minimum number of CEP rules. Therefore, the refinement process must provide a priority to select the most frequent state-values first to achieve the desired level with the minimum number of rules.

Based on the previous requirements, the following concepts illustrate the refinement process:

Let E be the event stream source.

Let V be the knowledge base that represents domain knowledge D.

Let $K = \{k_1, k_2, ..., k_n\}$ be a set of attributes in the knowledge base V.

Let k_i be the selected knowledge attribute where $k_i \in K$.

Let $V(k_i) = \{v_1, v_2, v_3, ..., v_n\}$ be a set of all extracted state-values for knowledge attribute k_i.

Let P be the original complex pattern specification.

Let P_{new} be the new complex pattern specification.

Let $W(v_i)$ be the number of complex events detected by pattern P and match state-value v_i.

Let t be the total number of complex events detected by pattern P.

Definition 4.6: *Value Coverage* $C(v_i)$. The ratio between the number of events that match a given state-value (v_i) to the total number of events detected by the original pattern P represented by t.

$$C(v_i) = W(v_i)/t \tag{4.2}$$

Let n be the total number of inferred state-values for the knowledge attribute k_i,

$$\therefore W(v_1) + W(v_2) + .. + W(v_n) = t$$

Thus, the overall coverage for the original pattern P is defined as:

$$C(P) = \sum_{i=1}^{n} C(v_i)$$
$$C(P) = \frac{W(v_1) + W(v_2) + .. + W(v_n)}{t} = 1$$

Referring to expression 4.1, the refined pattern coverage is defined as:

$$C(P_{new}) = \sum_{i=1}^{r} C(v_i) \tag{4.3}$$

where $1 \leqslant r < n, \quad 0 < C(P_{new}) < C(P)$.

The refined pattern coverage $C(P_{new})$ is determined by r. Accordingly, by using a complexity threshold to govern the value of r, the *SOI* can be represented with multi tradeoffs ranging from 1 to r.

Definition 4.7: *Pattern Complexity Threshold R_{max}.* is the maximum number of detection rules allowed to refine pattern P as set by the user.

Thus, $1 \leqslant r \leqslant R_{max}$

The R_{max} threshold is used to govern the level of complexity for pattern $C(P_{new})$ to meet refinement requirements in Section 4.2.1. Consequently, the targeted *SOI* can be represented with multi-tradeoffs $(1..R_{max})$.

Nevertheless, fulfilling refinement requirements in Section 4.2.2 requires using the opposite approach. That is, selecting a tradeoff value that produces the closest pattern coverage to the desired outcome. The refinement outcome is expressed by a *Minimum Support Threshold (MST)*.

Definition 4.8: *Minimum Support Threshold MST.* The ratio of new pattern coverage $C(P_{new})$ to original pattern coverage $C(P)$ that indicate the level of Situation Refinement required by the given scenario as set by the user.

That implies,

$$MST \leq C(P_{new}) < C(P)$$
$$\because SOI \subset S$$
$$\therefore SOI = f\left(S, D, MST\right)$$

The following is a typical Situation Refinement scenario based on the previous concepts.

Scenario 2: In a weather monitoring system, the user uses pattern P to detect temperature readings from all stations with temperature reading exceeds 58 F'.

P: **TempRead**(TimeStamp, StationID, Reading)
    WHERE (Reading>58);

Suppose that the user wishes to refine the existing situation by 60% (i.e., by setting MST value to 0.6) using the attribute *'Location'* in the *'Stations'* knowledge base to identify locations most affected by such a rise. In the given scenario, there are two input data sources: event stream *'TempRead'*, and knowledge base *'Stations'* which holds the current state-values for a given attribute *Location*.

Table 4.2 shows the complex event distribution by *Location* (listed in descending order), and their coverage value. The state-value labeled 'A1' is the most frequent state-value (the highest number of detected readings) with coverage = 0.501. Since this value is less than the Minimum Support Threshold as set by the user ($MST = $

Table 4.2: Complex event distribution by location
(in descending order)

Location	# of detected readings where temprature>58 F	$C(v_i)$
'A1'	501	0.501
'A4'	192	0.192
'A2'	190	0.190
'A11'	76	0.076
'A7'	41	0.041
Total	**1,000**	**1**

0.60). Hence, the successive frequent value 'A4' is accumulated to the previous one to satisfy that support threshold. Consequently, the Minimum Support Threshold is met since the overall coverage reaches (0.693),

$$C(P_{new}) = \sum_{i=1}^{r} C(v_i) = 0.501 + 0.192 = 0.693 \tag{4.4}$$

where $r =$ the number of state-values in $R(k)$.

Since both state-values ('A1', 'A4') produce the closest coverage that matches the MST threshold, therefore, both state-values are the minimal subset of state-values that satisfies the desired outcome.

Definition 4.9: *Frequent Values Set* $R(k)$. The minimal subset of state-values of $V(k)$, which delivers a combined coverage that satisfies the MST threshold.

Referring to Table 4.2 where $MST = 0.60$,

$\because V(\text{'Location'})=\{\text{'A1'}, \text{'A4'}, \text{'A2'}, \text{'A11'}, \text{'A7'}\}$

$\therefore R(\text{'Location'})=\{\text{'A1'}, \text{'A4'}\}$ where $R(k) \subset V(k)$

Accordingly, the refined pattern P_{new} is expressed as follows,

P_{new}: **TempRead**(T.Stamp, StationID, Reading)
      WHERE (Reading>58)
        AND (Location='A1' **OR** Location='A4')

To reduce pattern complexity, the refinement requirements in Section 4.2.1 implies selecting the minimal set of rules that generates the closest coverage to MST. To achieve such a closeness, the process of adding a new state-value into $R(k)$ must be done progressively, i.e., by gradually accumulating value coverage v_i for the most frequent state-values to $C(P_{new})$, then, adding the successive frequent state-value until the desired MST is reached. Section 4.4.2 illustrates the Gradual Inclusion process to identify the minimal number of frequent state-values.

4.4.2 Gradual Inclusion Process: Gain Versus Complexity

The Gradual Inclusion process aims at achieving the highest pattern coverage using minimal rule complexity, i.e., with the minimum number of state-values. Hence, it must give priority to the elements with the highest coverage first.

To achieve such a goal in a single scan, the state-values with the highest coverage must be gradually added to the $R(k)$ set until reaching an overall coverage $C(P_{new})$ that satisfies the MST threshold. To illustrate such a concept, Fig. 4.2 shows the

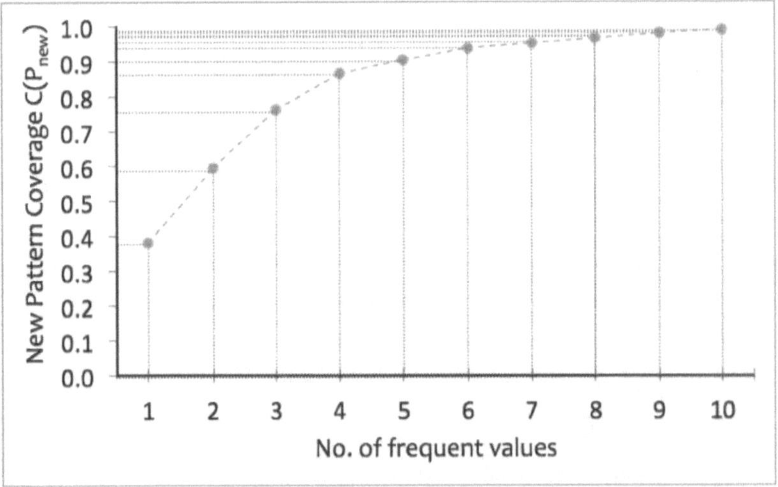

Figure 4.2: New pattern coverage vs. number of state-values

gradual increase in the refined pattern coverage $C(P_{new})$ as new *state-value* added to the $R(k)$ set.

Adding a state-values on a descending order causes an incremental decline in the amount of gained overall coverage $C(P_{new})$ as the number of state-values increased.

From the Situational Awareness view, such a decline has a negative impact; specifically, the amount of gained coverage from adding a state-value with a small coverage does not justify the increased pattern complexity resulted from such an addition.

Thus, the Gradual Inclusion must be restricted to the state-values higher than a particular threshold to reduce the cost of the increased pattern complexity in favor of small gains using a *Lower Bound Threshold*. In other words, it must be terminated as soon as $C(v_i)$ goes beyond that threshold.

Definition 4.10: *Lower Bound Threshold LBT*. this threshold represents the minimum coverage required by a single *state-value* to be considered frequent as set by the user.

Thus, $C(v_i) \geq LBT$

So far, two measures are used to govern the outcome of the Gradual Inclusion process: the LBT threshold to eliminate the state-values with lower coverage; and the MST threshold to terminate the inclusion process when the level of situation refinement is reached.

4.4.2.1 State-Values Similarity

Another issue must be examined when the MST threshold is used. Precisely, it is the case where two successive state-values (v_i, v_{i+1}) have a similar (very close) value coverage, yet, only state-value v_i is considered in the minimal set $R(k)$ because of the MST threshold.

In Table 4.2, for example, by applying MST=0.60, only state-values labeled 'A1' and 'A4' are selected in the minimal set with an overall coverage 0.693. The issue here is that both state-values 'A4' and 'A2' are too close in their coverage 0.192 and 0.190; yet, only 'A4' is selected in the minimal set.

As shown in Fig. 4.3, both 'A4' and 'A2' nearly imply the same significance from Situational Awareness perspective. Hence, the Gradual Inclusion process must have an exception to include 'A2' in the minimal set regardless of the MST restrictions.

One way to enforce such an exception is by assessing the difference between the two state-values using a similarity threshold. If that difference is less than the similarity threshold, then state-value v_{i+1} is added to the minimal set.

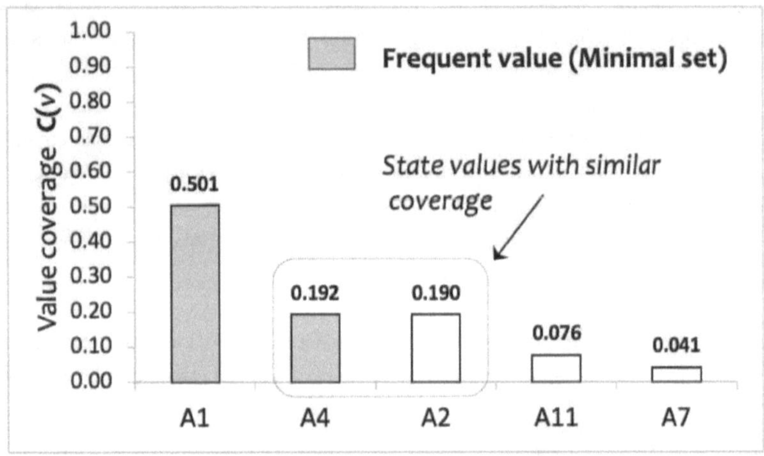

Figure 4.3: State-values similarity

Definition 4.11 *Similarity Threshold d_{max}*. This is the maximum acceptable difference in coverage between two successive state-values to be considered similar as set by the user,

where, $0 \leq d_{max} \leq 1$

Definition 4.12: *Normalized difference $d(v_i, v_{i+1})$*. The relative difference between two consecutive values (v_i, v_{i+1}) in their coverage is defined as,

$$d(v_i, v_{i+1}) = \frac{C(v_i) - C(v_{i+1})}{C(v_1) - C(v_{i+1})} \tag{4.5}$$

Hence, to consider both values to be similar,

$d(v_i, v_{i+1}) \leq d_{max}$

To sum up the Gradual Inclusion process, Fig. 4.4 illustrates how to enforce Situation Refinement requirements (Section 4.2.1 and 4.2.2) by using LBT, MST, and d_{max} thresholds to identify the minimum number of state-values.

Section 4.4.3 demonstrates how to maintain state-value coverage during the streaming period following the Situation Refinement requirement in Section 4.2.3.

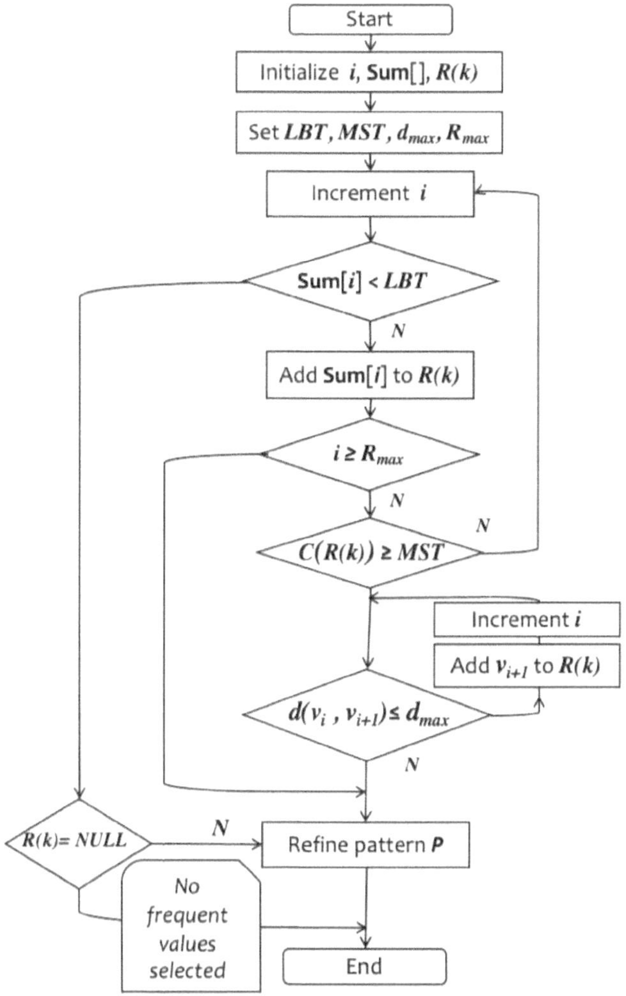

Figure 4.4: The Gradual Inclusion Process

4.4.3 Maintaining Value Coverage in the Event Stream

As per refinement requirements stated in Section 4.2.3, the number of detected complex events per time-window changes over time due to the dynamic nature of the event stream. Such dynamic distribution must be reflected in the *SOI* as the event stream evolves. This can be achieved by using a weight factor that reflects the period density in the Gradual Inclusion process.

In Fig. 4.5, for example, by observing two different time-windows $w(0{:}1)$ and $w(4{:}5)$, the number of detected complex events in the time-windows is 5 and 10, respectively. Also, the value count for a particular state-value v in these windows is 3 and 6. Accordingly, the value coverage in $w(0{:}1)$ is ($\frac{3}{5} = 0.6$), and in $w(4{:}5)$ is ($\frac{6}{10} = 0.6$). Although both time-windows have the same coverage, but the number of

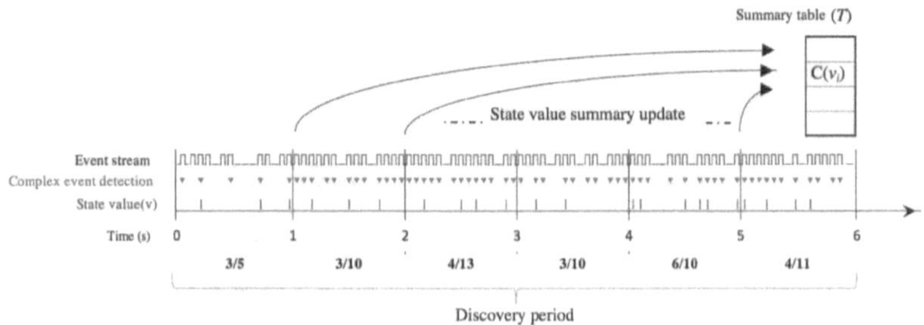

Figure 4.5: Summary update

complex events in these windows are different.

To represent the change in weight as the event stream evolves, two conditions must be met: First, only up-to-date value coverage (the stored summary) is stored in the memory regardless of the event stream size. and second, since the ratio between the elapsed time and the current time-window is changing over time, hence, both stored summary and the current value coverage must be weighted according to that ratio before the stored summary is updated.

To satisfy these conditions, a Relative Weight factor (λ_n) is introduced to redistribute the overall value coverage between the stored value in the summary table T and the current time-window. The relative weight of the current window (λ_n) is:

$$\lambda_n = \frac{W_n}{\sum_{i=1}^{n} W_i} \tag{4.6}$$

where,

W_n denotes the number of detected complex events in the current window,

W_i denotes the number of detected complex events for window i.

At the end of each time-window, λ_n is used to re-distribute the overall coverage between the stored summary and the new state-value coverage as follows:

$$T(v) = T(v)(1 - \lambda_n) + C(v)\lambda_n \tag{4.7}$$

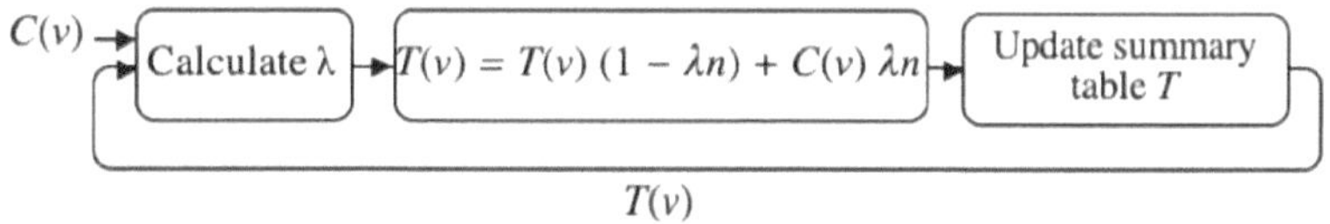

Figure 4.6: Value coverage re-distribution over time

Fig. 4.6. shows the continuous update of the summary table T and the value coverage re-distribution as the event stream evolves. Table 4.3. tabulates the weight re-distribution over time for the event stream sample in Fig. 4.5.

Table 4.3: Complex event detection over time

Time window (n)	Complex events per window (W)	Complex events so far (t)	The Relative weight (λ)	
			Current window	Past history
1	5	5	1.000	0.000
2	10	15	0.667	0.333
3	13	28	0.464	0.536
4	10	38	0.263	0.737
5	10	48	0.208	0.792
6	11	59	0.186	0.814

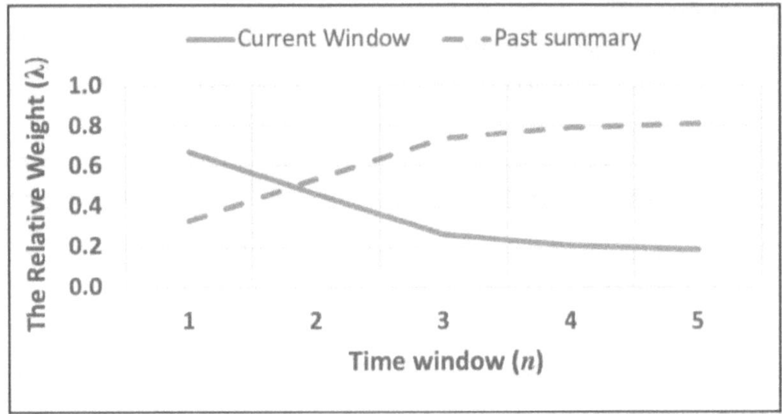

Figure 4.7: Relative Weight re-distribution over time

As the event stream evolves, the weight assigned to the stored summary $(\lambda - 1)$ increases in favor of the current window coverage (λ); simultaneously, the contribution of current window coverage to the overall summary becomes less notable as the ratio between both periods increases as shown in Fig. 4.7.

To sum up, the relative weight re-distribution over time ensures two things: First, it maintains the continuous weight re-distribution between past event history and the current time-window as the event stream evolves, and second, a fixed amount of memory is used as only the last summary is kept in the summary table T at any given time.

The next section illustrates the various Situation Refinement Model components along with the related tasks. The Summary Update Algorithm is detailed in Section 4.5.3.

4.5 The SRM Model: Components, Architecture, and Tasks

In this Section, we present the components of the Situation Refinement Model: Section 4.5.1 exhibits the proposed model architecture, and the interaction between various model components. Section 4.5.2 illustrates Situation Refinement tasks and triggers. Section 4.5.3 shows the Summary Update Algorithm to re-distribute value coverage over time.

4.5.1 Model Architecture

As shown in Fig. 4.8, the Situation Refinement Model architecture consists of three main components: the CEP engine where complex event detection takes place, the External Knowledge which provides state-values for knowledge attribute k, and the Rule Refinement Module where the selection of the minimal set of state-values takes place.

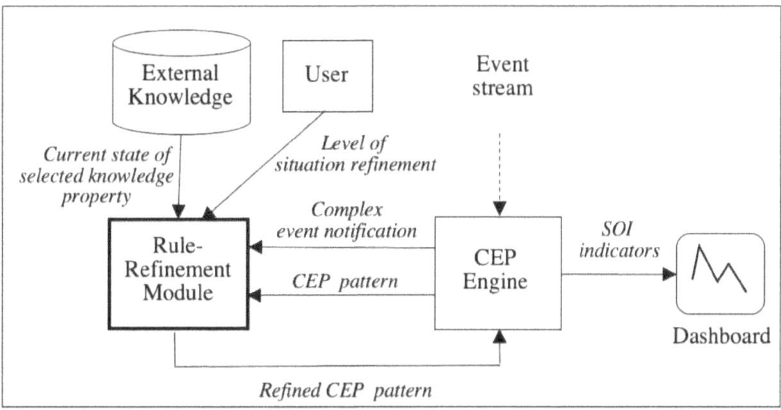

Figure 4.8: Situation Refinement Model Architecture

4.5.2 The Situation Refinement Model Tasks

Table 4.4 shows Situation Refinement Model tasks. The system initiates each task according to four triggers, as shown in Fig. 4.9. The task description is as follows:

Table 4.4: Situation Refinement Task Discription

No	Trigger	Task
1	Start of situation discovery <End user>	**Detection**: CEP engine continuously monitors events stream and detects complex events that match pattern P.
2	Complex event detected < CEP engine>	**Acquisition**: Rule Refinement Module acquires state-values from the knowledge source.
3	End of time-window <Timer>	**Summary update**: Rule Refinement Module updates state-values occurrences in the summary table.
4	End of situation discovery <End user>	**Refinement**: Rule Refinement Module activates the Gradual Inclusion process.

1. **Detection:** The user initiates the refinement process by triggering the start of situation discovery. Next, the CEP engine starts detecting the occurrences of complex events in the event stream using the original complex event pattern P.

2. **Acquisition:** During complex event detection, the Rule Refinement Module acquires the current state-values for the selected knowledge attribute k from the external knowledge source.

3. **Summary Update:** At the end of each time-window, the Rule Refinement Module updates the acquired state-values summary in the summary table T.

4. **Refinement:** After the user request, the Rule Refinement Module ends the situation discovery period and activates the Gradual Inclusion Process.

4.5.3   Summary Update Algorithm

Algorithm 1 shows the summary update for the acquired state-values at the end of each time-window. The discription of Algorithm 1 is as follows:

1. The system initializes the summary update parameters (lines 1:9).

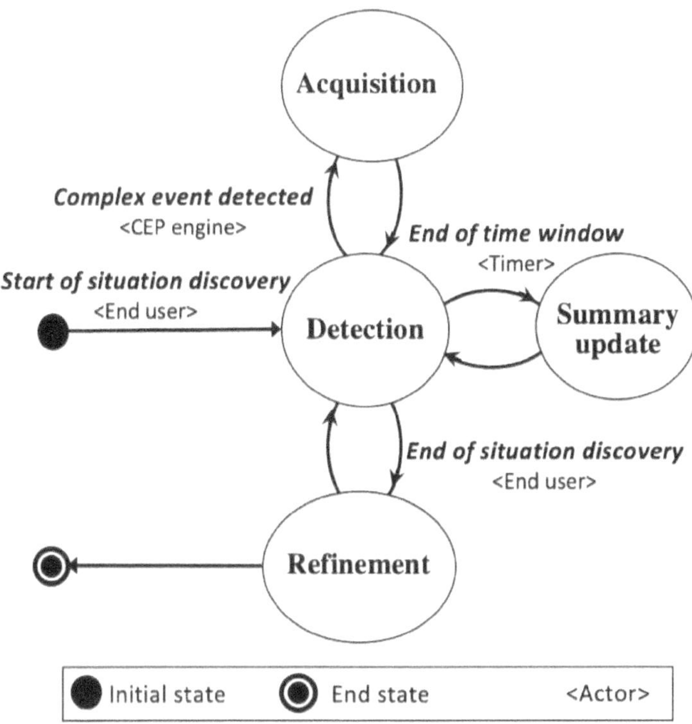

Figure 4.9: Situation Refinement Model: the state diagram

2. The system initializes the CEP engine and event stream (lines 10:11).

3. The system waits for user trigger to start situation discovery (line 12).

4. Upon complex event detection, the state-values statistics are updated with new state-value occurrences (lines:13-21).

5. At the end of each time-window, the system calculates the relative weight factor λ (line:22).

6. Based on the value of λ, the system calculates and inserts the new state-value into the summary table (lines:23-31).

Algorithm 1: Summary Update

Input: Event stream S, Complex event pattern P, External knowledge K
Output: Summary table Sum
1: Let Ce be the complex event detected by P
2: Let v be the current state-value for Ce in K
3: Let C be the current coverage for v
4: Let V be the value occurrence table
5: Let Sum be the summary table
6: Let W be the event counter in current window
7: Let t be the number of complex events detected
8: Let λ be the relative weight for current window
9: Let N be the current window number
10: Initialize CEP engine (P)
11: Initialize event stream S
12: WAIT for start of Situation Discovery
13: **while** NOT end of Situation Discovery **do**
14:    **while** NOT end of current window **do**
15:      **if** Ce is detected **then**
16:        Increment event counter W
17:        Increment total event counter t
18:        Acquire v for Ce
19:        Increment value occurrence for v in V
20:      **end if**
21:    **end while**
22:    $\lambda = W/t$
23:    **for all** v IN V **do**
24:      $v = V(value)$
25:      $C = V(occurrence)$
26:      Lookup v IN Sum
27:      **if** NOT exist **then**
28:        Insert v in Sum
29:      **end if**
30:      $sum(v) = (sum(v) * \lambda - 1) + (C * \lambda_i)$
31:    **end for**
32:    Reset table V
33:    Reset event counter W
34:    Increment window counter N
35: **end while**

In the next section, both the Summary Update and the Refinement tasks are demonstrated. Also, the impact of changing time-based settings such as window size and the duration of the discovery period on the SOI outcome are examined.

4.6 Investigating the Impact of Changing Situation Refinement Parameters

4.6.1 Experimental Work Description

In Section 4.4, parameters LBT, R_{max}, D_{max}, and MST were investigated, formulated, and demonstrated with several examples. Though, the impact of changing the value of time-based parameters such as window size W and the duration of the discovery period on the SOI outcome still requires further experimental investigation using a time-stamped dataset.

Therefore, this section aims to investigate the impact of changing window size on the pattern complexity and pattern coverage, as well as the impact of the duration discovery period length on the SOI outcome. Hence, the following relations are examined:

- The relationship between window size W and pattern complexity.

- The relationship between window size W and overall coverage $C(R)$.

- The impact of discovery period's length on the overall situation discovery.

To formally conduct experimental work here, these relations are formalized in hypotheses as follows:

4.6.2 Test Hypotheses

4.6.2.1 Hypotheses 1 and 2

Purpose: To test the effect of changing window size in both pattern complexity and overall coverage $C(R)$.

Hypothesis 1: *"Changing window size with fixed minimum support threshold MST will change the number of selected state-values; consequently, pattern complexity will be affected"*.

H_0: $Count(R(k)_{old}) = Count(R(k)_{new})$
H_1: $Count(R(k)_{old}) \neq Count(R(k)_{new})$

Hypothesis 2: *"Expanding window size will increase overall coverage $C(R)$"*.

H_0: $C(R)_{old} = C(R)_{new}$

H_1: $C(R)_{old} > C(R)_{new}$
H_2: $C(R)_{old} < C(R)_{new}$

Description: To satisfy the test requirements, MST is set at fixed values while the window size is changing. The window size is set to four different sizes (1, 2, 3, and $4s$) to examine the impact of the window size variable on the pattern complexity and pattern coverage. The test is repeated with different discovery periods to ensure the consistency of the test results. The discovery period is incremented by $10s$ at every iteration.

4.6.2.2 Hypothesis 3

Purpose: To test the impact of increasing the discovery period size on the overall situation discovery.

Hypothesis 3: *"Expanding discovery period has no significant effect on overall situation discovery"*.

H_0: $R_{old} = R_{new}$
H_1: $R_{old} \neq R_{new}$

Description: To satisfy test requirements, the whole dataset is divided into ten fixed time intervals ($3s$ each). At each time interval, a relative weight (λ) for both the past period and the current window is observed.

4.6.3 The Test Environment Setup

Fig. 4.10 shows various components of the test package for this experimental work. The test package contains three components: *Esper* CEP engine, the main test module, and the input datasets.

The main test module composed of two sub-modules: the *Summary Update* module that performs the summary update process as exhibited by Algorithm 1 and the *Gradual Inclusion* module as exhibited in Fig. 4.4.

Two input datasets are used during the test: First, an event dataset that contains time-stamped events depicting various patient activities [157], and second, the knowledge dataset that holds *state-values* that discribe related patient activity [158].

Both the *Summary Update* and the *Gradual Inclusion* modules have been implemented using **R** [159]. For the CEP engine module, the *Esperr* CEP [160], an event-based analytics package, is used to detect complex events during the streaming period. Fig. 4.10 shows the four-step process to test each hypothesis:

1. Test setup: The system user sets the input test variables (MST, W, LBT, R_{max}), and the number of test iterations.

2. System startup: The system user initializes *Esper* CEP engine and the input event stream.

3. Summary update: where a complex event is detected, the *Summary Update* module acquires the state-value from the knowledge dataset and updates the summary table T.

4. Refinement: At the end of the situation discovery period, the *Gradual Inclusion* Module provides the minimal set of state-values $R(k)$.

Since the test hypotheses requirements imply executing refinement process with different discovery period and time-window settings, the test execution time is set as shown in Table 4.5. It should be noted that the actual response time (after terminating the discovery period) is measured in the PC machine used in this test (Intel Core i5, 2.3 GHz).

Table 4.5: Test execution time settings

Hypothesis	# of test iterations	Discovery period under test (sec)	Time-Window under test (sec)	Test streaming time (sec)	Actual response time per iteration (ms)
1	12	10, 20, 30	1, 2, 3, 4	240	≈ 3
2	12	10, 20, 30	1, 2, 3, 4	240	≈ 3
3	10	3, 6, 9, ..., 30	4	165	≈ 3

Throughout this test, the state-values for the knowledge attribute '*Description*' are presented as symbolic alphabets (*A*, *B*, *C*, etc.) to make test results more readable. Table. 4.6 specifies the test settings used.

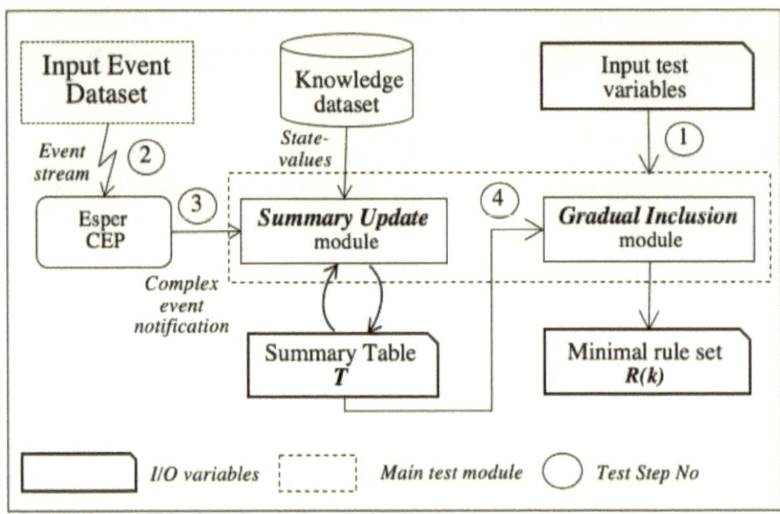

Figure 4.10: The test workflow diagram

Table 4.6: Test settings

Item	Discription
Event dataset	*Patient data set representing 30 days' record log. Fields: Date, Time and Code*
Dataset size	2004 event records.
External knowledge dataset	*A CSV file contains description for various patient activities. Fields: Code, Description.*
No. of state-values	20
Testing suite	RStudio, Version 1.1.456 [161]
CEP Engine	*Esperr* for Event-based Analytics [160, 162]

4.6.4 Test Results

After conducting hypotheses testing with input variables as indicated in the previous Section, the following results are obtained:

1. *Hypothesis* 1: Table 4.7 shows the state-values coverage versus window size for discovery periods ($10s$, $20s$, and $30s$). Fig. 4.11 shows the change in the selected number of state-values (pattern complexity) as the window size increases. The state-values (blue color) are frequent as they fulfill the MST threshold (red bar). As seen in Fig. 4.11, the change in window size affects the number of selected state-values. Specifically, as the window size increases, the number of frequent state-values is reduced. This reduction is due to the gradual increase in coverage against state-values with higher weights against values with smaller weights. This can be seen in Table 4.7 as the most frequent values A and J (highlighted in yellow) gain increased coverage against the least frequent ones (D, H) when expanding window size from $3s$ to $4s$.

Such impact is less notable in large discovery periods ($20s$ and $30s$) as most state-value gain stable coverage regardless of the change in window size.

Overall, the window size change affects the number of frequent values needed to satisfy the MST threshold. Fig. 4.11 shows such an impact as the number of frequent values drops when the window size is changed to $3s$ as the case with state-values B and G (Fig. 4.11.a) and state-value D (Fig. 4.11.b).

Thus, the null hypothesis H_0 is rejected, and the alternate hypothesis H_1 is accepted; therefore,

"Changing window size with fixed minimum support threshold MST will change the number of selected state-values; consequently, pattern complexity will be affected".

H_0: $Count(R(k)_{old}) = Count(R(k)_{new})$ ×
H_1: $Count(R(k)_{old}) \neq Count(R(k)_{new})$ ✓

2. *Hypothesis* 2: The impact of changing window size on overall coverage $C(R)$ can be seen in Fig. 4.12. As the window size increases, the overall coverage decreases regardless of the stream size. The decrease can be observed with different samples under test in Table 4.7. Hence, the null hypothesis H_0 is rejected, and the alternate hypothesis H_1 is rejected, while hypothesis H_2 is accepted; therefore,

"Expanding window size decreases the overall coverage $C(R)$".

H_0: $C(R)_{old} = C(R)_{new}$ ×
H_1: $C(R)_{old} > C(R)_{new}$ ×
H_2: $C(R)_{old} < C(R)_{new}$ ✓

Table 4.7: State-values coverage vs. window Size

Discovery period (s)	State value (v)	Window size (s)			
		1	2	3	4
		Value coverage $C(v)$			
10	A	0.4216	0.4216	0.4291	0.4343
	B	0.1453	0.1453	0.1415	0.1400
	G	0.1395	0.1395	0.1372	0.1353
	J	0.1338	0.1338	0.1459	0.1450
	D	0.1046	0.1046	0.1033	0.1031
	H	0.0548	0.0548	0.0000	0.0000
20	A	0.4149	0.4149	0.4149	0.4149
	B	0.1447	0.1447	0.1447	0.1446
	G	0.1399	0.1399	0.1399	0.1399
	J	0.1224	0.1224	0.1224	0.1224
	D	0.1049	0.1049	0.1049	0.1050
	H	0.0732	0.0732	0.0732	0.0731
30	A	0.4177	0.4177	0.4149	0.4149
	B	0.1452	0.1452	0.1447	0.1446
	G	0.1422	0.1422	0.1399	0.1399
	J	0.1230	0.1230	0.1224	0.1224
	D	0.1008	0.1008	0.1049	0.1050
	H	0.0712	0.0712	0.0732	0.0731

3. *Hypothesis* 3: Fig. 4.13 shows the change in the relative weight over time. As can be noted from the simple moving average (*SMA*) line in the chart, there is a steady decline in the relative weight value throughout the successive periods in the test. Therefore, the impact of relative weight on the overall coverage becomes less noticeable as the stream evolves. The reason behind such decline is the increased deviation between past weight $\lambda(t_0 : t_{n-1})$ and the relative weight for the current time-window $\lambda(n)$.

In summary, as the discovery period expands, the impact of any recent change in state-value coverage on the overall situation discovery is minimal. Thus, the null hypothesis H_0 is accepted, while the alternate hypothesis H_1 is rejected; therefore,

"Expanding discovery period has no significant effect in overall situation discovery".

H_0: $R_{old} = R_{new}$ ✓
H_1: $R_{old} \neq R_{new}$ ✗

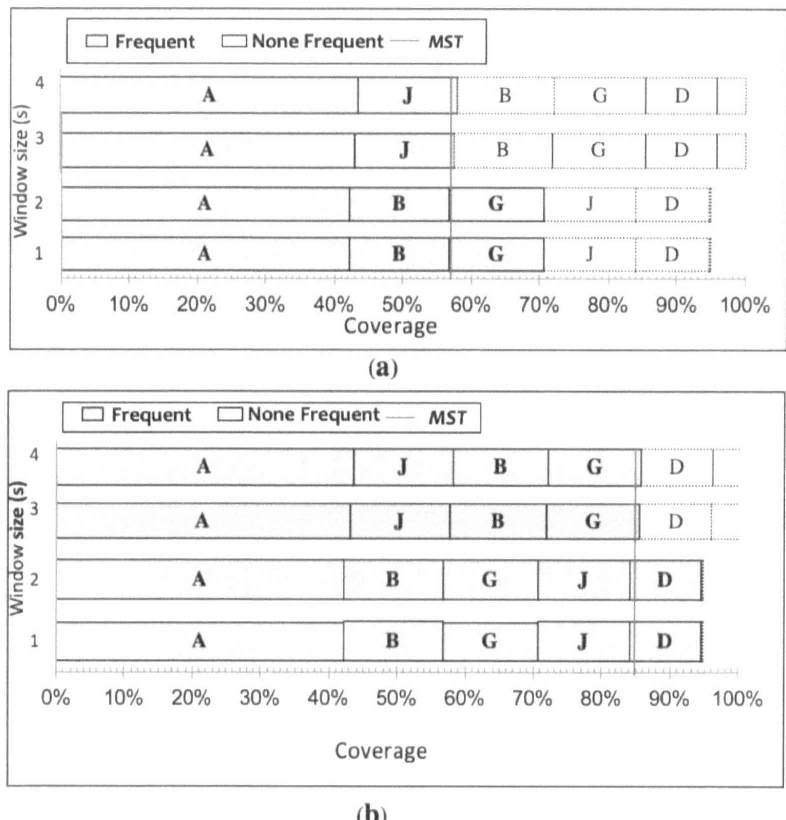

Figure 4.11: Window size vs. pattern complexity

4.6.5 Results Summary

Based on the experimental work results, the following findings are observed:

• Although the refinement requirements presented in Section 4.2 propose refining the existing *SOI* using the minimal set of detection rules; however, both requirements differ on their targeted outcome. In requirement 4.2.1, for example, the system users want to attain the flexibility of viewing the targeted *SOI* with different representations (tradeoffs). In contrast, requirements 4.2.2 implies identifying the closest tradeoff that

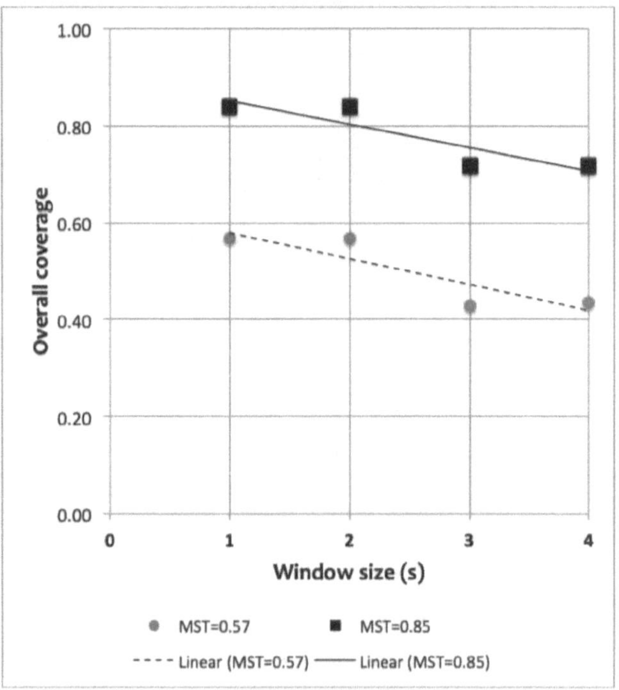

Figure 4.12: Overall coverage vs. window size

meets with the required level of situation refinement MST. Therefore, to improve the overall model performance against various refinement scenarios, the scenario type must be incorporated in the **SRM** model.

• As seen by hypotheses 1 and 2, the window size is significant to pattern complexity and coverage. Therefore, the window size must be taken into account when evaluating the refinement outcome.

• As observed in hypothesis 3, the relative weight stability over a long duration has an adverse impact on the refinement outcome since the recent trends have a minimal effect on the overall situation discovery. Therefore, identifying the minimum period for the situation discovery is essential in resolving such a pitfall. One possible approach to mitigate this problem is by observing the downtrend of the SMA in Fig.

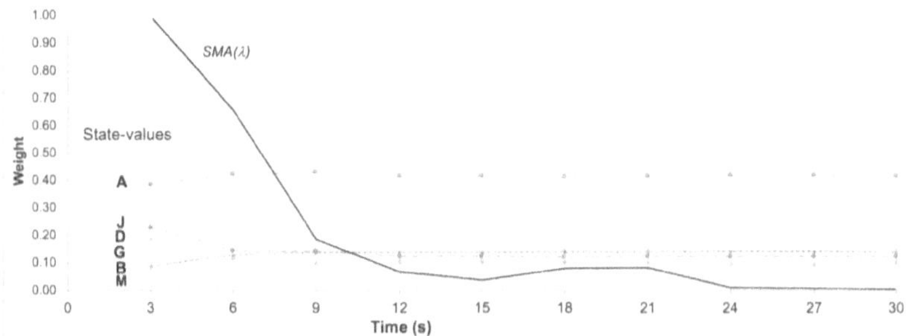

Figure 4.13:   Relative weight change over time

4.13. In other words, the Situation Refinement process should be terminated as soon as the change becomes less noticeable.

4.7 Chapter Summary

In this Chapter, we presented a CEP-based Situation Refinement model (**SRM**) to refine the existing detection rules towards identifying a particular *SOI*.

The **SRM** fulfills the **book** contributions No.1 and 2, as follows:

• To achieve the highest refinement outcome with the minimum pattern complexity, **SRM** uses the *Gradual Inclusion Process* (Section 4.4.2), which uses the *LBT* and *MST* thresholds to accumulate state-values in descending order and in a single scan mode. The user-preferred context is represented by the current state (state-value) of the selected Knowledge attribute E.

• To tackle the issue of increased mined elements in an unbounded event stream, **SRM** uses the Relative Weight factor (λ_n) during the Summary Update (Section 4.5.3) to re-distribute the overall coverage between the past stored coverage and the current time-window.

So far, the **SRM** delivers the targeted pattern specification P_{new} based on a user-defined thresholds (MST, R_{max}). Both thresholds govern the outcome of the **SRM** concerning the targeted gain (pattern coverage) and the required cost (pattern complexity).

Although the **SRM** delivers the minimum number of detection rules required to refine a particular situation S, yet, the refinement outcome still depends on a user-defined setting. That is, as these setting changes, the number of detection rules changes accordingly. Hence, the following question needs to be addressed:

-For a given refinement scenario, what are the MST and R_{max} values that deliver the best possible refinement outcome?

Toward addressing this issue, in the next Chapter we present a cost-gain tradeoff measure to identify cost and gain settings that deliver the best possible refinement outcome for a given refinement scenario.

Chapter 5

Assessing Cost and Gain Tradeoff in Situation Refinement for Complex Event Proccessing

In this Chapter, we present a cost-gain tradeoff measure to deliver the best possible tradeoff concerning refinement cost and gain for **SRM**. The proposed measure aligns with the **book** contributions #3 and #4 (Section 1.1).

5.1 Introduction

CEP systems rely on detection rules based on complex pattern queries to filter, correlate, and aggregate events in a near real-time manner. Recognizing emerging situations in near real-time requires expressive pattern definitions capable of detecting and matching events based on reasoning over event streams, and the available domain knowledge.

In CEP terms, the expressiveness means the ability to express CEP detection rules with a high-level representation. This could be achieved by identifying CEP detection rules based on the correlation between event data and domain knowledge [12,163,164].

CEP systems typically enrich pattern expressiveness by using matching techniques such as windowing, sequencing, conjunction, and disjunction [7]. Consequently, the complexity of CEP pattern queries increases. Further, the increased pattern expressiveness leads to a more complex situation representation that is challenging to analyze. Moreover, it increases the computational cost associated with matching complex pattern queries with the incoming stream of events [20, 26].

Although the expressive CEP pattern queries require more complex detection rules to derive high-level information, such complexity is needed in many emerging application areas to recognize situations of interest and improve situational awareness accurately [165, 166].

Towards minimizing the impact of increased pattern complexity, several CEP related studies have focused either on reducing the number of CEP detection rules to maximize throughput and minimize latency [31, 50, 62, 167] or optimizing CEP query plans with available computational resources [168, 169]. Other research efforts have focused on developing cost models to estimate the cost of using computational resources while running CEP queries [166, 168, 170].

There exist several cost estimation methods within the data mining community, such as decile analysis [171], lift charts [172], cumulative lift charts, and cumulative gains [173]. Though, none of the existing CEP cost methods focus on addressing pattern complexity issues by considering the other side of the problem, that is, the gain concerning additional perceived knowledge as a result of adding new detection rules to the existing CEP pattern. Therefore, identifying the best possible tradeoff between cost and gain is essential to deliver the resultant SOI at the proper representation level.

The remainder of this Chapter is organized as follows: Section 2 presents existing work related to the CEP system's cost evaluation methods. The problem background and the proposed solution are presented in section 3. Section 4 illustrates the process of deriving the proposed tradeoff measure for the **SRM**. In section 5, the Cost-Gain evaluation algorithm is used to determine the best possible tradeoff using various refinement curve characteristics. In section 6, conclusions are made.

5.2   Situation Refinement: Cost and Gain Tradeoff

As stated in Section 4.7, the refinement outcome of the **SRM** depends on the user-defined setting, which implies a different number of detection rules each time this setting is changed; consequently, the resultant SOI will have a different presentation. To demonstrate this, The following pattern template P presents a pattern specification used to define a particular situation S, as illustrated in Scenario 1:

P : **Select** Time, SensorID, TempCelsius, GeoArea

    From TempReading.win:time(500 minutes)

    Where (TempCelsius > 30)

Scenario 1: An observatory research center uses a scattered network of sensors to deliver temperature data for scientific research purposes. The original complex pattern P is used to detect the number of rises in temperature greater than 30 celsius within 500 minute period (Situation discovery period). The temperature readings streamed to the CEP engine every 10 minutes. Table 5.1 shows a sample of CEP notifications generated during the given period.

Table 5.1: CEP notifications sample

Time	SensorID	TempCelsius	GeoArea
02:31:29	S101123	31	A1
02:41:29	S101824	33	A1
02:41:29	S102007	36	B3
02:51:29	S103040	34	A2
	.	....	....
	.	....	....
04:11:29	S104001	33	C2
04:31:29	S101131	31	A1

The system users aim to improve their knowledge about the given situation by identifying a new *SOI* through situation refinement. For example, the system users might use the geographical area attribute '*GeoArea*' to obtain more focused insights about geographic areas that are mostly affected by such a rise in temperature.

Let P_{ini} be the original pattern used to define the original situation S.

Let P_{new} be the new pattern used to identify the resultant *SOI*.

Let *SOI* be the refined representation of situation S.

Let k be the selected knowledge attribute.

Let n be the total number of *state-values* of k.

Let $V(k) = \{v_1, v_2, v_3, ..., v_n\}$ be the set that holds *state-values* of k.

Let $R(k) = \{v_1, v_2, v_3, ..., v_i\}$ be the set of unique frequent state-values and $R(k) \subset V(k)$.

Let r_m be an additional detection rule where r_m is $(k = v_m)$ and $1 \leq m \leq i$.

Let $C(v_m)$ be the coverage that denotes the number of events that match a *state-value* v_m over the total number of events detected by the original pattern P_{ini}.

Table 5.2 shows the **SRM** output summary for complex events detected by pattern P_{ini} in Scenario 1. The *state-values* of attribute *GeoArea* are listed as ordered by the GIP, i.e., based on their coverage $C(v)$ in descending order. The following cases show the refinement outcome for Scenario 1 under different *MST* settings.

Table 5.2: **SRM** output summary

GeoArea	# of detected readings	Coverage $C(v)$
'B1'	189	0.750
'A2'	28	0.111
'A1'	10	0.040
'C5'	7	0.028
'B3'	5	0.020
'D1'	5	0.020
'A4'	3	0.012
'E2'	2	0.008
'A3'	2	0.008
'C2'	1	0.004
Total	252	1

- **CASE #1:** The user sets the *MST* threshold to 0.71. As a result, the GIP selects the *state-value* 'B1' since its coverage satisfies the *MST* threshold; hence, the resultant pattern is expressed as follows:

P_1: **Select** Time, SensorID, TempCelsius, GeoArea

    From TempReading.win:time(500 minutes)

    Where (TempCelsius > 30)

    AND (*GeoArea*= 'B1')

Therefore, the corresponding resultant *SOI* generated is:

SOI_1: "75.00 % of the detected readings originated in **B1** area".

- **CASE #2:** The user sets the *MST* threshold to 0.90. As a result, the GIP selects *state-values* 'B1', 'A2', and 'A1' to form the new detection rule since their combined coverage satisfies the *MST* threshold $(0.750 + 0.111 + 0.040 = 0.901)$; hence, the resultant pattern is expressed as follows:

P_2: **Select** Time, SensorID, TempCelsius, GeoArea

    From TempReading.win:time(500 minutes)

    Where (TempCelsius > 30)

    AND (*GeoArea*= 'B1' **OR** *GeoArea*= 'A2' **OR** *GeoArea*= 'A1')

As a result, the corresponding resultant *SOI* becomes:

SOI_2: "90.01 % of the detected readings are originated in either **B1** or **A2** or **A1** area".

5.2.1 Problem Statement

As seen in Scenario 1, increasing the number of detection rules leads to more pattern coverage (gain); however, it also leads to more complexity (cost). Hence, it is desirable to have a performance evaluation measure to determine the best achievable tradeoff between cost and gain while considering the given refinement scenario's particular settings.

There are two issues associated with the development of a performance evaluation measure: first, how to identify a reference refinement curve (cost vs. gain) to be used as a reference to evaluate the actual refinement curve while considering the unique characteristics of the two refinement modes (*Situation Discovery* and *Incremental Refinement*), and second, how to process the proposed performance evaluation in a single scan to comply with event streaming restrictions with respect to processing time and memory storage.

5.3 Related Work

In section 5.3.1, we provide a description of the notable work in current literature that focuses on the impact of pattern complexity on CEP query processing cost. Section 5.3.2 introduces various methods proposed to monitor and evaluate the cost of running CEP queries. In section 5.3.3, we present a cumulative gains chart, the cost evaluation tool under focus in this chapter, and the challenges that must be overcome to use it for event streaming scenarios.

5.3.1 Complexity Verses Processing Cost

Several studies have addressed the significance of evaluating the complexity issue of complex pattern queries and its impact on the overall performance of CEP systems [166,169,174]. These studies reveal the relationship between pattern complexity and the increased processing cost of CEP queries.

The *Cayuga* model [175] shows the gradual increase in the event stream throughput as the complexity of CEP query increases. Also, the *NFA^b* model [174] shows the linear relationship between the processing cost and the complexity of CEP query. Zhang et al. [166] use the theory of *Descriptive Complexity* [176] to emphasize the relationship between the expressive power of CEP query language and pattern complexity. Other studies review the linear relationship between the number of CEP rules and the impact of the operator type (Filter, Logical, Temporal) on the processing cost of CEP queries with respect to time [177–179].

Since the increased cost of processing CEP queries negatively impacts the overall system performance, several CEP cost evaluation methods use various metrics to monitor the cost of running CEP queries, such as CPU and memory consumption [180], CEP pattern matching cost [166], and the cost of query execution plan [181].

These studies show the significance and importance of developing CEP cost models to estimate the computation cost required to execute CEP queries or to select the optimal plan to distribute the workload among available computation resources. The next subsection highlights recent main contributions in CEP cost models.

5.3.2 CEP Cost Models

A range of CEP cost models has been proposed to estimate and monitor the cost of processing CEP queries. Akdere et al. [182] present a plan-based cost model to select the optimal plan for event detection using communication cost and minimum latency. Schultz-Møller et al. [183] introduce the *Next CEP* system, which includes a cost model to re-write and optimize the CEP query. Teymourian et al. [184] present a semantic enrichment model to select the optimal processing plan based on the cost of the event and background knowledge acquisition. There are several other CEP cost models proposed to select the optimal execution plan based on CPU and communication cost [168,185–187].

Although the cost models mentioned above show the importance of query rewriting and plan distribution techniques in gaining a higher system throughput, they only focus on estimating the computation and communication costs related to processing CEP while ignoring pattern complexity as a cost factor.

Many other cost models focus on selecting the optimal query plan based on the cost of processing CEP operators, such as the CEP cost model in [188], the *ZStream* model [168], the analytical cost model in [170], and the weight-based cost model proposed in [189]. Also, the study by Zhang et al. [166] leverages the theory of *Descriptive Complexity* [176] by categorizing pattern query features into different types of complexity classes.

In general, the existing cost methods focus on evaluating CEP performance in terms of throughput, latency, and memory consumption while ignoring the impact of increasing pattern complexity on the knowledge representation layer of *SOI*.

Subsection 5.3.3 illustrates various statistical methods used to assess the performance of a given model. In particular, the cumulative gains chart which is the most preferred evaluation tool, as noted by the DMA Research Council Group [171] and others [190, 191].

5.3.3   Model Performance Evalution Methods

In data mining community, several statistical methods are used to assess the performance of a given model such as the Kolmogorov-Smirnov (K-S) statistics [192, 193], decile analysis [171, 194, 195], lift charts [171, 172], and cumulative gains chart [173, 190, 193]. These methods aim at identifying the model that generates the highest gain with the least cost [171]. In general, they follow two approaches: evaluating more than one competing model at the same time, or assessing one model only by including cost evaluation metrics in the data mining process [193, 196].

In its original form, the cumulative gains chart is an evaluation tool to assess a model's performance by showing how well the given model performs under different cost settings.

Like any other evaluation metrics oriented to static data types, the classical cumulative gains chart suffers from a major drawback in applications with data streams as this metric requires sorting the given dataset before calculating Area Under Curve (AUC) for the given model. Thus, this approach can not be applied directly in data streams where all computation must be performed in a single scan mode [197, 198].

With that regard, several efforts have successfully addressed the issue of calculating the AUC in the evolving data streams, such as the incremental learning framework [199], the Prequential AUC algorithm [198, 200], and the AUC estimation algorithm [201].

However, using AUC to assess cost of **SRM** refinement requires taking into account additional unique constraints: First, in **SRM**, the process of the cost evaluation must be applied to a particular zone in the chart (Refinement Zone) rather than the whole AUC in the chart; second, the cost metric must consider both **SRM** refinement modes as they imply different requirements [28]. Specifically, the diagonal line in the cumulative gains chart represents the worst model performance in the *Situation Discovery* mode. In contrast, the same line represents the ideal performance in the *Incremental Refinement* mode.

Section 5.4 presents the problem definition and the proposed solution to tackle the challenges mentioned above.

5.4   Problem Background

To evaluate cost and gain of a situation refinement, a reference curve is needed to assess the given refinement curve.

Definition 5.1: *Reference refinement curve*: the curve that represents the maximum refinement outcome concerning cost and gain in a particular refinement mode.

The refinement cost r denotes the number of detection rules required to achieve a particular gain with $1 \leq r \leq R_{max}$, where R_{max} is the maximum acceptable refinement cost. Refinement gain G denotes the combined pattern coverage with a valid range of $0 \leq G \leq 1$. $G=0$ corresponding to no refinement gain and $G=1$ denotes the maximum refinement gain.

Definition 5.2: *Actual refinement curve*: the curve that represents the actual outcome of situation refinement concerning cost and gain in a particular refinement mode.

Definition 5.3: *Refinement Zone (RZ)*: the area in the cost-gain chart bounded by the minimum refinement cost ($r=1$), maximum acceptable refinement cost ($r=R_{max}$), minimum acceptable gain ($G= MST$), and maximum gain ($G=1$). Both the R_{max} and MST thresholds are defined by the user. The RZ area defines the boundaries where the actual curve performance is evaluated, as shown in Fig. 5.1.

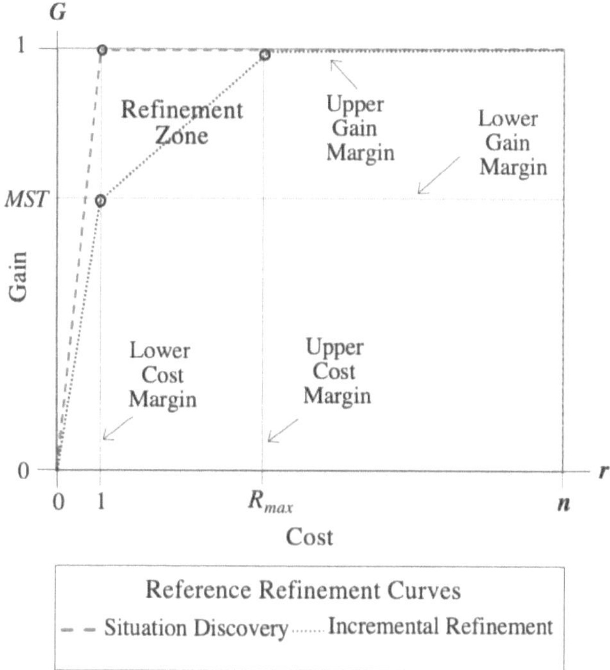

Figure 5.1: Reference curves for *SD* and *IR* modes

5.4.1 Visualizing Reference Refinement Curves

Fig. 5.1 shows the reference refinement curves for both refinement modes based on the requirements mentioned above. The vertical axis (G) denotes the refinement gain. The horizontal axis (r) denotes the refinement cost. The red dashed curve depicts the reference refinement curve in the *SD* mode. The blue dotted curve represents the reference refinement curve in the *IR* mode. The gray area in the graph represents the Refinement Zone (*RZ*) with the following cost-gain margins:

- The Lower-Cost Margin is at $r=1$.
- The Upper-Cost Margin is at $r=R_{max}$.
- The Lower-Gain Margin is at $G=MST$.
- The Upper-Gain Margin is at $G=1$.

Both refinement modes imply different requirements with respect to the refinement outcome. In the *Situation Discovery* mode (*SD* mode), the user seeks to reach the maximum refinement gain at the lowest cost. That implies the reference refinement curve in the best case scenario reaches the highest gain ($C{=}1$) with minimum cost at $r{=}1$, as shown by the red dashed line in Fig. 5.1.

Different than the *SD* mode, the *Incremental Refinement* mode (*IR* mode) imposes a different refinement target. Since most complex events require a response in near real-time, the resources allocated to respond must be sufficient to react to these complex events at any given time. Hence, the *IR* mode is used for scenarios when resources are limited with respect to the number of complex events detected [28]. Since the overall pattern coverage (gain) denotes the maximum number of complex events that the system can handle at any given time, in the *IR* mode the user aims to respond to a limited number of complex events based on the available resources. For example, the number of fire trucks available versus the number of complex events detected that indicates a fire occurrence.

As the number of complex events is governed by pattern coverage (gain), a user seeks to perform such governance by using the refinement cost r. That is, as r increases, the refinement gain increases, and vise versa. Specifically, the user's objective is to have the flexibility to increase or decrease the gain between the minimum acceptable gain at $r{=}1$ to the maximum achievable gain at $r{=}R_{max}$. In order to control the resources efficiently, the increase in gain at each increment of r should be equal, therefore the acceptable refinement gain range ($MST \leq G \leq 1$) should be evenly distributed among the number of rules available for refinement ($r{=}1$ to R_{max}).

Given that, the reference curve in the *IR* mode must have the following characteristics: first, it must intercept points ($G{=}MST$, $r{=}1$) and ($G{=}1$, $r{=}R_{max}$), and second, the increase in gain at each increment of r must be equal. In other words, the *IR* mode's reference curve must exhibit a linear increase within the *RZ* area as shown by the blue dotted line in Fig. 5.1.

5.4.2 Actual Curve Performance Evaluation

In general, cost and gain evaluation methods need a reference curve to evaluate the actual curve. That is, as the difference in gain above or below the reference curve is minimized, the better the performance is delivered.

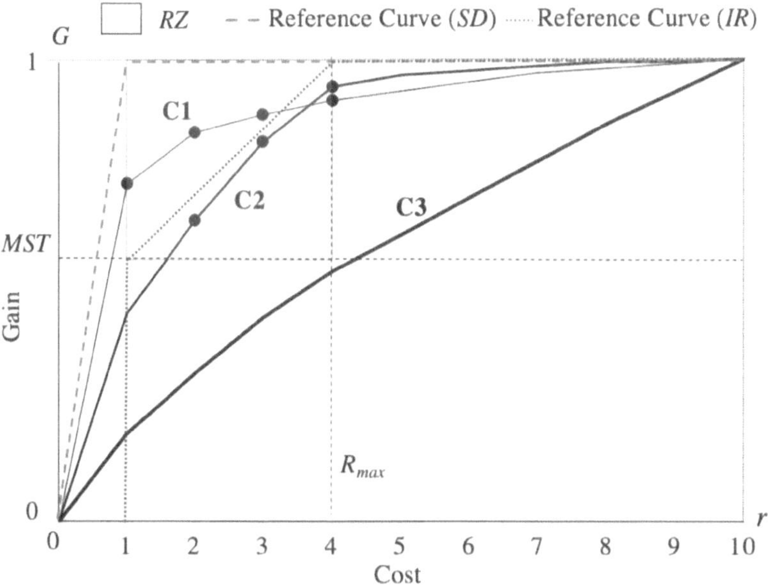

Figure 5.2: Refinement cost vs. refinement gain

To visually demonstrate this concept, Fig.5.2 shows three actual refinement curves C1, C2, and C3 and the reference curves for the two refinement modes. By using the *SD* curve as a reference curve (red dashed curve), it is clear that curve C1 provides better performance as compared to C2 since it is closer to the reference curve. By using the *IR* curve (blue dotted curve) as a reference curve, curve C2 performs better than curve C1 since it is closer to the reference curve. C3 represents a zero performance in both *SD* and *IR* modes as compared with C1 and C2 since it is completely out of the *RZ* boundaries.

To develop a performance measure for **SRM** based on the concept mentioned above, the following three unique **SRM** restrictions need to be addressed:

1. Contrary to the traditional cost evaluation methods where a single reference curve is used to assess the actual curve performance, the **SRM** has two re-finement modes (*Situation Discovery* and *Incremental Refinement*), each with its own reference curve. We propose a performance evaluation measure that is

sensitive to both refinement modes simultaneously.

2. While the evaluation process in the cumulative gains chart considers all points on the actual curve, the evaluation process in the **SRM** must only consider points within the user defined thresholds (i.e., within the RZ zone). Since there are changes based on the refinement scenario's particular settings, reference curves within the RZ boundaries for both refinement modes must be dynamically adapted to such change.

3. In the traditional cost evaluation methods, all data points of the actual curve must be stored in memory before the evaluation process. Therefore, the evaluation process in such a case is performed in two passes: first, acquiring actual data points, and second, evaluating these data points. However, the actual curve coordinates in the **SRM** must be delivered in near real-time to avoid data queuing and memory storage as the event stream evolves. The proposed evaluation measure accommodates this restriction by acquiring and evaluating actual curve data points in a single scan during the Gradual Inclusion Process (GIP).

In Section 5.5, we present a performance measure to evaluate the actual refinement curve while considering the restrictions mentioned above. The performance measure will then be used to determine the cost-gain values that deliver the highest refinement performance for the two refinement modes simultaneously.

5.5 Deriving the SRM Performance Measure

The SRM unique requirements imply using different performance measure for each refinement mode (IR and SD). Subsections 5.5.1 and 5.5.2 illustrate the derivation of the IR performance measure and the SD performance measure, respectively. First, some necessary definitions:

Definition 5.4: *Valid tradeoff point*: Any point $p(cost, gain)$ on the actual refinement curve within the RZ area representing the actual gain G at a particular cost r.

Definition 5.5: *Actual curve performance η*: The value that represents the performance of the actual refinement curve relative to the reference curve for valid tradeoff points in the RZ area where $0 \leq \eta \leq 1$. $\eta{=}0$ denotes the lowest performance score (zero performance) where the given curve is completely out of the RZ area. The higher the η is, the better performance is provided by the actual curve. $\eta{=}1$ denoted the highest performance, where both the actual and the reference curve are identical, and the given scenario is the perfect candidate for situation refinement.

5.5.1 Deriving Performance of Actual Refinement Curve: *IR* Mode

Fig. 5.3 shows two actual refinement curves C1 and C2, and the reference curve for the *IR* mode (blue dotted curve). The black dots in the actual curve denote valid tradeoff points (within the *RZ* area). The red dots denote invalid tradeoff points (outside of the *RZ* area).

Let i be the r index where $1 \leq r \leq R_{max}$.
Let η_i be the performance score for point p_i where $1 \leq i \leq R_{max}$.
Let $\eta(IR)$ be the actual curve performance where $\eta(IR) = \sum_{i=1}^{R_{max}} \eta_i$.
Let b_i be the absolute difference between the actual and the reference curve at $r{=}i$.

Since the tradeoff points of the actual curve might situate above or below the *IR* reference curve, the proposed performance measure computes each point's performance score in the actual curve (η_i) below or above the *MST* threshold, and then accumulates all η_i scores to η.

According to Fig. 5.3,
1) Since the highest $\eta(IR)$ value is 1, that implies each valid point above the *MST* threshold can have a maximum η_i score of $\frac{1}{R_{max}}$.
2) The highest η_i achieved when both points are identical; that is, the absolute difference (b_i) is 0. The absolute difference b_i is computed as follows:

Let a_i be the reference gain above the *MST* threshold at $r{=}i$.
Let $\triangle Gm_i$ be the actual gain above the *MST* threshold at $r{=}i$.
Let $\beta{=}1\text{-}MST$ be the maximum absolute difference between the actual curve and the reference curve in the *RZ* area.

Accordingly,
$$b_i = |\triangle Gm_i - a_i|$$

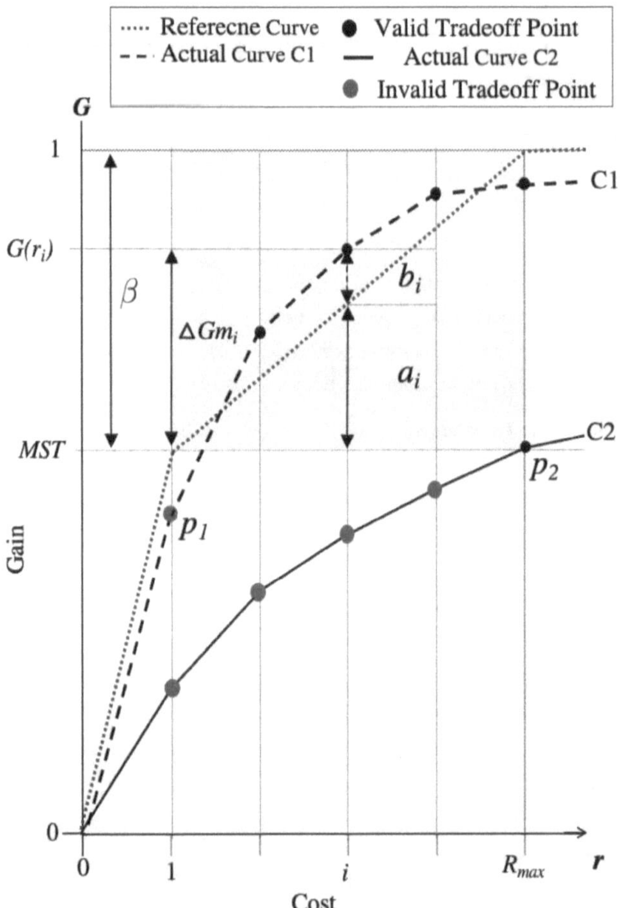

Figure 5.3: Actual curve evaluation (*IR* mode).

Therefore, any point p_i has a reference gain a_i as follows:

$$\frac{\beta}{R_{max} - 1} = \frac{a_i}{i - 1}$$

$$a_i = \frac{\beta(i - 1)}{R_{max} - 1}$$

3) **Valid point evaluation**: An actual curve delivers the highest performance score if $\sum_{i=1}^{R_{max}} \eta_i = 1$, that implies the maximum m_i score of any valid point p_i is $\frac{1}{R_{max}}$. Such a case occurs if the b_i value of point p_i is 0. Also, a given valid point p_i's performance score is $\eta_i=0$ if $b_i = \beta$, as shown in Fig. 5.3. Accordingly, η_i of any valid point is expressed as follows:

$$\eta_i = \left(1 - \frac{b_i}{\beta}\right) \times \frac{1}{R_{max}}$$

$$\eta_i = \left(\frac{\beta - b_i}{\beta}\right) / R_{max}$$

$$\eta_i = 1 - \frac{\beta - b_i}{R_{max}\beta} \qquad where \ \ 0 \leq \eta_i \leq \frac{1}{R_{max}} \tag{5.1}$$

4) **Invalid point evaluation**: Since points below the MST threshold are invalid, the η_i score of such points must have no contribution to $\eta(IR)$ regardless of their closeness to the reference curve. That is, $\eta_i=0$.

5) According to 3 and 4, the proposed performance measure assesses valid and the invalid point as follows:

• Case 1: if all the points in the actual curve are valid, all points in the RZ area will contribute to $\eta(IR)$ with η_i value > 0.

• Case 2: if the actual curve has valid and invalid points, the valid points will contribute to $\eta(IR)$ with η_i value > 0, whereas the invalid points with $\eta_i=0$.

• Case 3: if the actual curve is situated completely below the MST threshold, all invalid points will contribute with $\eta_i=0$ and, in turn, $\eta(IR)$ will be 0.

6) According to expression 5.1, the actual curve performance is expressed as follows:

$$\eta(IR) = \sum_{i=1}^{R_{max}} 1 - \frac{\beta - b_i}{R_{max}\beta} \qquad where \ \ 0 \leq \eta(IR) \leq 1 \tag{5.2}$$

5.5.2 Deriving Performance of Actual Refinement Curve: *SD* Mode

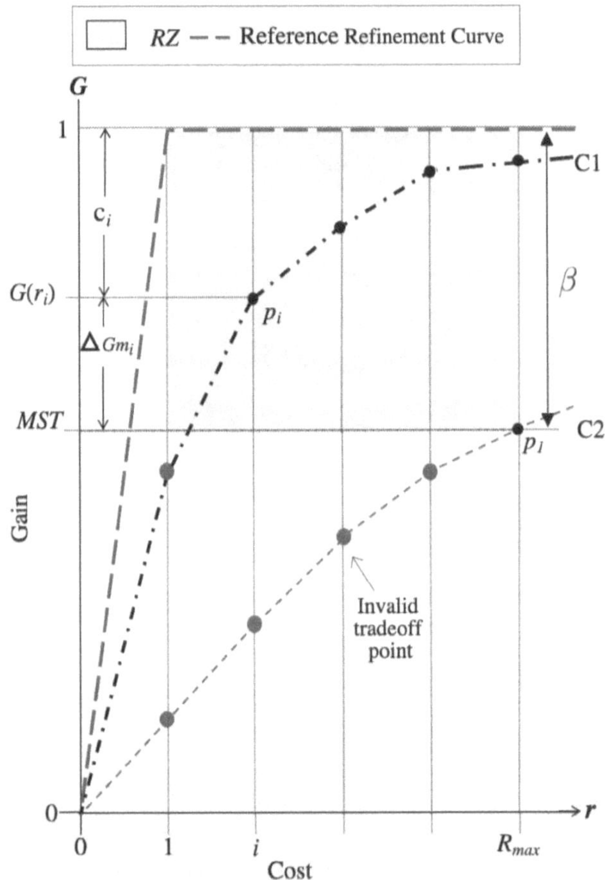

Figure 5.4: Actual curve evaluation (*SD* mode)

Fig. 5.4 shows two actual refinement curves C1 and C2, and the reference curve in the *SD* mode. Given that the reference curve in the *SD* mode reaches the maximum gain (G=1) at r=1, the actual curve becomes identical to the reference only if all valid tradeoff points reach the maximum gain (G=1). Based on that, the actual curve performance can be derived as follow:

Let $\eta(SD)$ be the actual curve performance and $0 \leq \eta(SD) \leq 1$.
Let p_i be a valid tradeoff point in the actual curve at r=i.
Let $G(r_i)$ be the actual gain at r=i.
Let $\triangle Gm_i$ be the gain above the MST threshold at r=i.
Let c_i be the difference in gain between the reference and actual curve at $r = i$.
Let c be the sum of all c_i's where $c = \sum_{i=1}^{R_{max}} c_i$.

1) **Valid point evaluation**: a given point p_i (Fig. 5.4) delivers the maximum performance if the difference in gain (c_i) is 0. In such a case, the c_i value can be expressed as follows:

$$c_i = 1 - (\triangle Gm_i + MST) = 0 \tag{5.3}$$

Also, a valid point p_i's zero performance occurs when c_i=1-MST, as the case with point p_1 (Fig. 5.4).

2) **Invalid point evaluation**: as per Def. 5.5, only valid tradeoff points contribute to the actual curve evaluation. Therefore, the c_i value of any invalid point p_i is substituted with 1-MST regardless of its closeness to the reference curve.

3) Based on 1 and 2, the c_i value for both valid and invalid points can be expressed as follows:

$$c_i = 1 - \left(\frac{\triangle Gm_i + |\triangle Gm_i|}{2} + MST \right) \tag{5.4}$$

4) Given that both the reference and actual curves are identical if c_i=0 for all the points where $1 \leq r \leq R_{max}$, the maximum curve performance occurs if the following condition is satisfied:

$$c = \sum_{i=1}^{R_{max}} c_i = 0$$

5) Based on 2, the actual curve zero performance occurs if all tradeoff points where

$1 \leq r \leq R_{max}$ are invalid (c_i=1-MST). In such a case,

$$c = \sum_{i=1}^{R_{max}} 1 - MST = R_{max}\beta$$

Accodingly, the performance of the actual curve in the SD mode is defined as:

$$\eta(SD) = 1 - \frac{c}{R_{max}\beta} \qquad where \;\; 0 \leq \eta(SD) \leq 1 \qquad (5.5)$$

By deriving the performance of the actual refinement curve, the η value can be used to determine the best cost-gain tradeoff T_{best}(cost, gain) for both refinement modes. In Section 5.6, this process is demonstrated using different refinement curves.

5.6 Determining the Best Cost-Gain Tradeoff

This section demonstrates how the derived performance measure η is used to determine the best tradeoff point $T_{best}(cost, gain)$ for both refinement modes.

Definition 5.6: *Best tradeoff* T_{best}: A tradeoff point in the actual refinement curve, which delivers the highest possible performance with respect to cost and gain, given a particular refinement setting (R_{max}, MST). To determine T_{best} in a particular refinement mode, the Cost-Gain Evaluation algorithm (Algorithm 2) is used to identify the r value that delivers the highest refinement performance. The evaluation algorithm complies with the **SRM** requirements as it accumulates and updates η value for both refinement modes in a single scan. Additionally, we investigate the impact of increasing the refinement gain on the actual curve performance.

5.6.1 Evaluation Process Setup

Fig. 5.5 shows the test workflow for the evaluation process. The evaluation test is conducted as follows:

1. Three syntactic datasets that represent various refinement curve characteristics are used to identify T_{best} for both refinement modes, as shown in Fig. 5.6.

2. The T_{best} is determined by gradually increasing the r value from 1 to R_{max}. At each r cycle, the η parameters ($G(r)_i, a_i, b_i, c_i, \triangle Gm_i$) are acquired, computed, and accumulated in a single scan (line #: 6-16).

Algorithm 2: Cost-Gain evaluation

Input: Dataset DS, MST, R_{max}
Output: $T_{best}(cost, gain)$

1   **Let** $\alpha \leftarrow R_{max} - 1$
2   **Let** $\beta \leftarrow 1 - MST$
3   **Let** $\eta(IR)_{max}, \eta(SD)_{max} \leftarrow 0$
4   **Let** $b, c \leftarrow 0$

    // Start (R_{max}) iterations
5     $i = 1$ **while** $i \leq R_{max}$ **do**
6      $Gr_i \leftarrow DS[i]$
7      $\triangle Gm_i \leftarrow Gr_i - MST$

     // Calculate additional gain for IR mode
8      $a_i \leftarrow (i-1)\beta/\alpha$

9   **if** $Gr_i > MST$ **then**
10       $b_i \leftarrow |\triangle Gm_i - a_i|$
11   **else**
12       $b_i \leftarrow \beta$
13   **end**

14      $b \leftarrow b + b_i$

     // Calculate difference in gain for SD mode
15      $c_i \leftarrow 1 - ((\triangle Gm_i + |\triangle Gm_i|)/2 + MST)$
16      $c \leftarrow c + c_i$

     // Calculate η for both modes
17      $\eta(IR) \leftarrow 1 - b/\alpha\beta$
18      $\eta(SD) \leftarrow 1 - c/R_{max}\beta$

     // Update best tradeoff point (T_{best})
19   **if** $\eta(IR) > \eta(IR)_{max}$ **then**
20       $\eta(IR)_{max} \leftarrow \eta(IR)$
21       $T_{best}(IR) \leftarrow (i, MST)$
22   **end**
23   **if** $\eta(IR) > \eta(SD)_{max}$ **then**
24       $\eta(SD)_{max} \leftarrow \eta(SD)$
25       $T_{best}(SD) \leftarrow (i, MST)$
26   **end**

     // Plot $\eta(IR), \eta(SD)$ in the chart
27   **Plot** $\eta(SD), \eta(IR)$
28      $i++$
29 **end**
     // Output best tradeoff point
30   **Output** $T_{best}(SD), T_{best}(IR)$

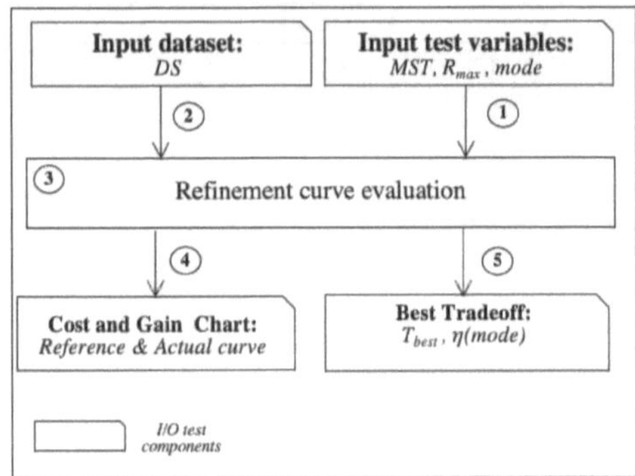

Figure 5.5: The test workflow diagram

3. The evaluation algorithm examines and updates the highest η reached so far, and it is corresponding T_{best} at each increase in r (line #: 17-26).

4. As the r value reaches R_{max}, the test delivers T_{best} that produces the highest η in both refinement modes (line #: 30).

5. To observe the impact of increasing refinement gain on the performance of the refinement curve, steps 2, 3, and 4 are repeated three times, each with a different MST setting (0.3, 0.5, and 0.7).

5.6.2 Test Results

5.6.2.1 Test Results: Dataset (DS1)

Figures 5.7, 5.8, and 5.9 show the test results for dataset DS1 under different MST settings. Fig. 5.7 shows DS1 test results for MST=0.3, while Fig. 5.8 and 5.9 show the test results for MST=0.5 and MST =0.7, respectively. Based on the previous figures, the following are concluded:

• In the SD mode, the evaluation test identified T_{best} at (4, 0.3) as it delivers the highest performance with $\eta = 0.902$ (Fig. 5.7.a).

• In the IR mode, the test identified T_{best} at (3, 0.7) as it delivers the highest performance with $\eta = 0.411$ (Fig. 5.9.b).

• By observing curve performance under various MST settings (Fig. 5.10), the increase in the MST threshold from 0.3 to 0.5, and 0.7 leads to a slight increase in

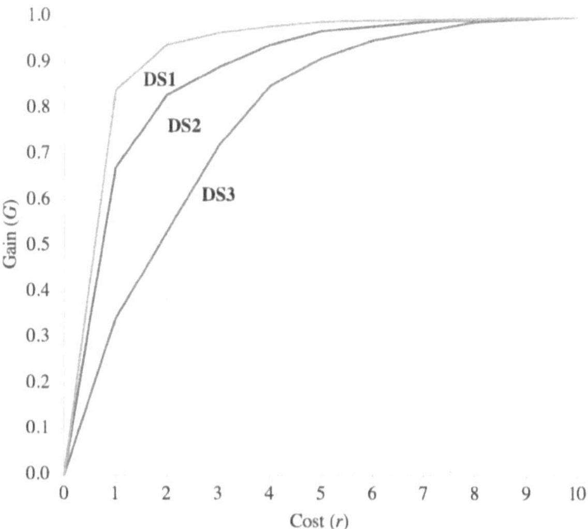

Figure 5.6: Actual refinement curves for datasets under test

curve performance in the IR mode, whereas it decreases in the SD mode.

• Despite such an increase in the IR mode, yet this curve characteristic shows an outstanding performance in the SD mode since it becomes closer to the upper-left-corner in the RZ space in all MST settings. In particular, the given curve characteristic is suitable for SD scenarios that require a lower MST value since it provides the highest performance $\eta(SD)=0.902$ at $MST=0.3$.

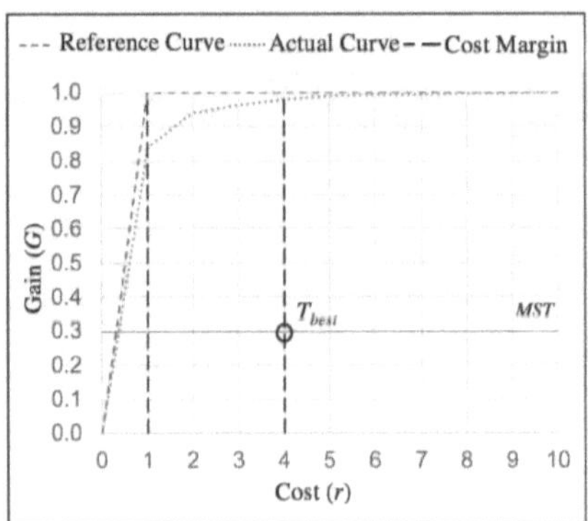

(a) SD mode: $\eta(SD)$=0.902 at R_{max}=4

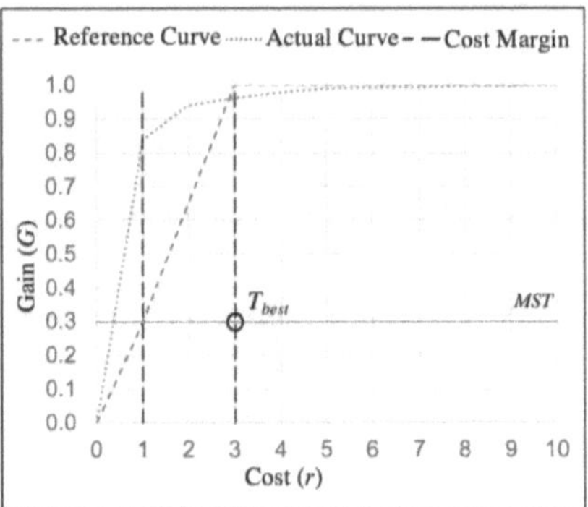

(b) IR mode: $\eta(IR)$=0.176 at R_{max}=3

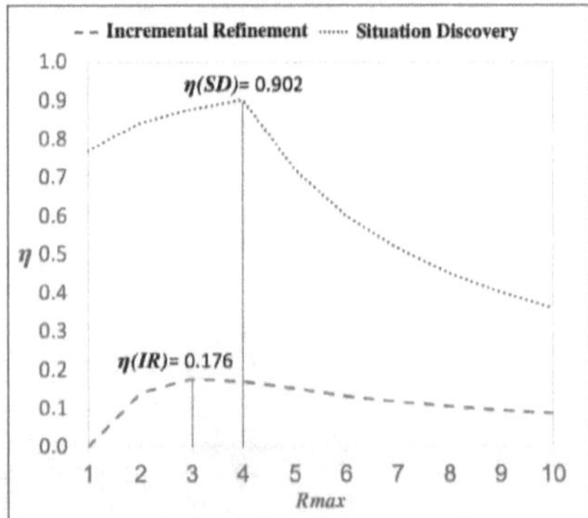

(c) The overall model performance.

Figure 5.7: Performance evaluation test: DS1, MST=0.3

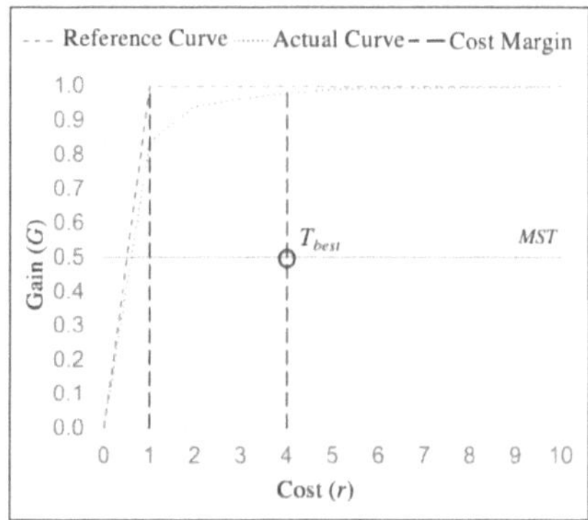

(a) *SD* mode: $\eta(SD)$=0.863 at R_{max}=4

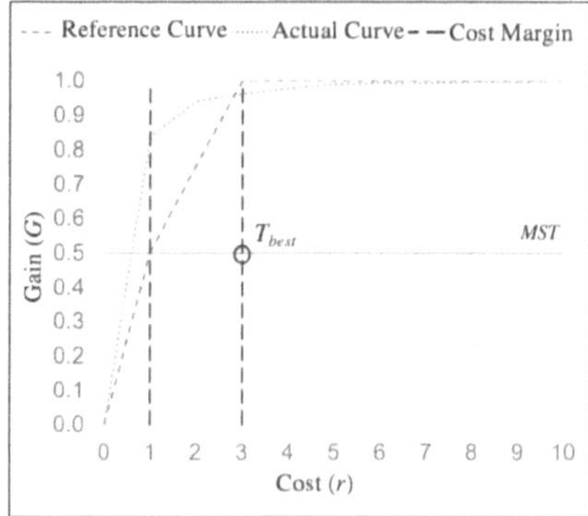

(b) *IR* mode: $\eta(IR)$=0.247 at R_{max}=3

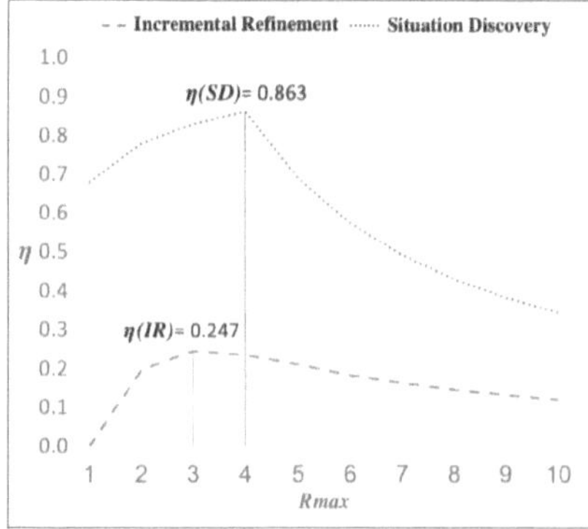

(c) The overall model performance.

Figure 5.8: Performance evaluation test: DS1, MST=0.5

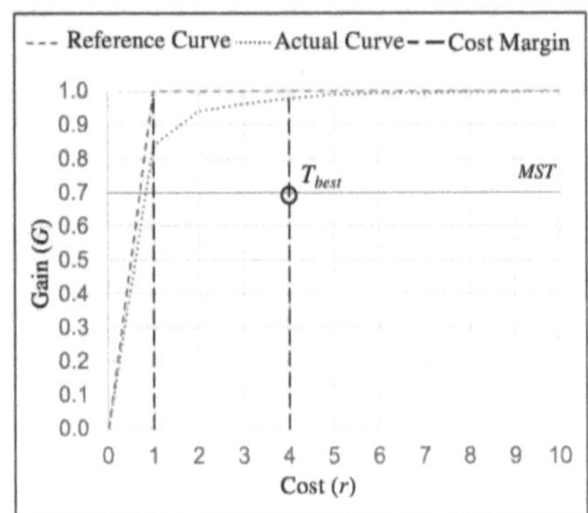

(a) *SD* mode: $\eta(SD)$=0.771 at R_{max}=4

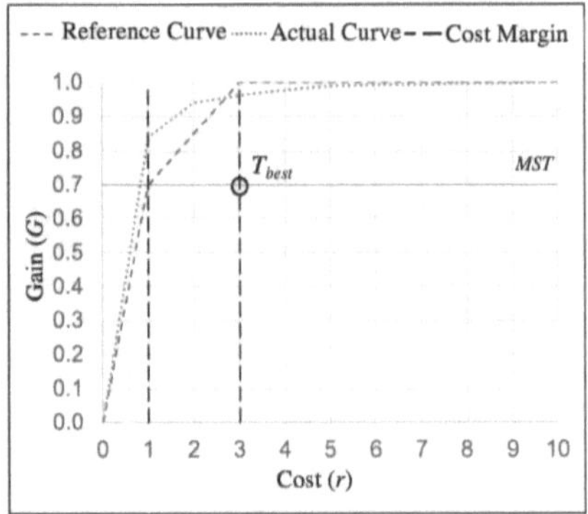

(b) *IR* mode: $\eta(IR)$=0.411 at R_{max}=3

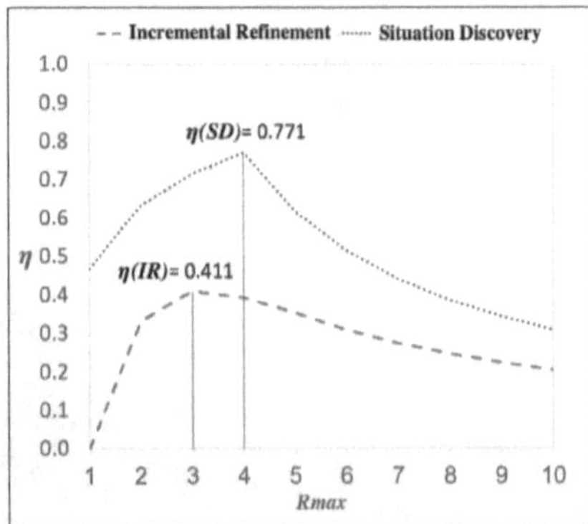

(c) The overall model performance.

Figure 5.9: Performance evaluation test: DS1, *MST*=0.7

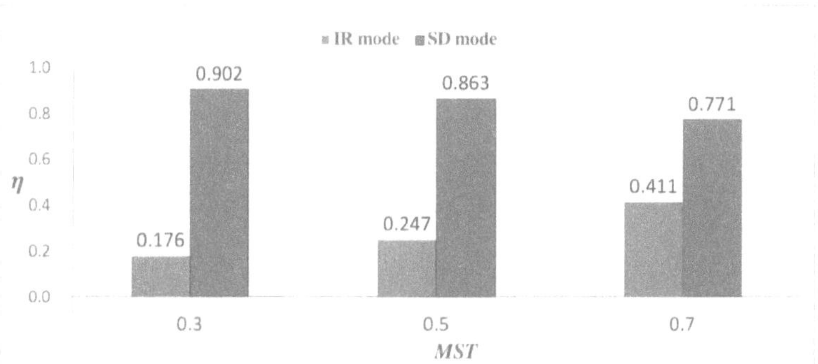

Figure 5.10: DS1: Actual Curve performance under various MST settings

5.6.2.2   Test results: Dataset (DS2)

Figures 5.11, 5.12, and 5.13 show the test results for dataset DS2 under different MST settings. Fig. 5.11 shows DS2 test results for MST=0.3, while Fig. 5.12 and 5.13 show the test results for MST=0.5 and MST =0.7, respectively. Based on the test results, the following are concluded:

• In the SD mode, the evaluation test identified T_{best} at (5, 0.3) as it delivers the highest performance with $\eta = 0.800$ (Fig.]5.11.a).

• In the IR mode, the evaluation test identified T_{best} at (4, 0.7) as it delivers the highest performance with $\eta = 0.833$ (Fig. 5.13.b).

• As shown in Fig. 5.14, increasing the MST threshold from 0.3 to 0.5 and 0.7, the curve shows a significant decline in performance in the SD mode. On the contrary, the MST threshold increase has a positive impact on the IR mode since the $\eta(IR)$ reaches 0.833 at MST=0.7.

• Overall, the given curve characteristics represent a perfect candidate for IR scenarios that require a higher MST setting. Besides, the given curve is suitable for the SD scenarios that require a lower MST setting.

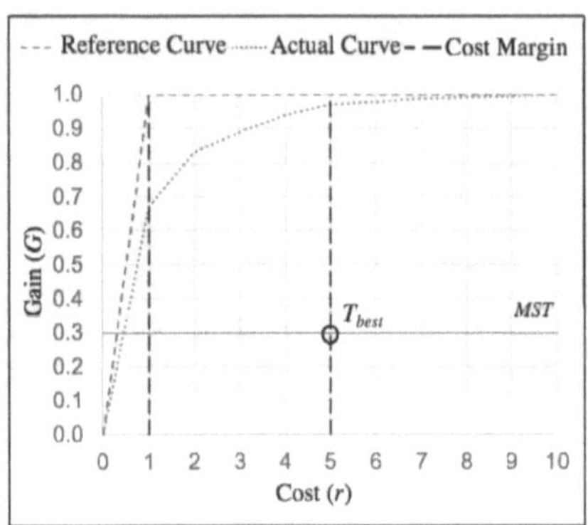

(a) *SD* mode: $\eta(SD)$=0.800 at R_{max}=5

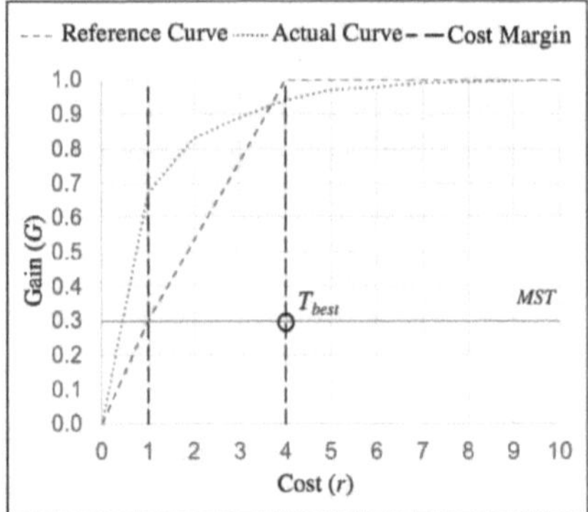

(b) *IR* mode: $\eta(IR)$=0.393 at R_{max}=4

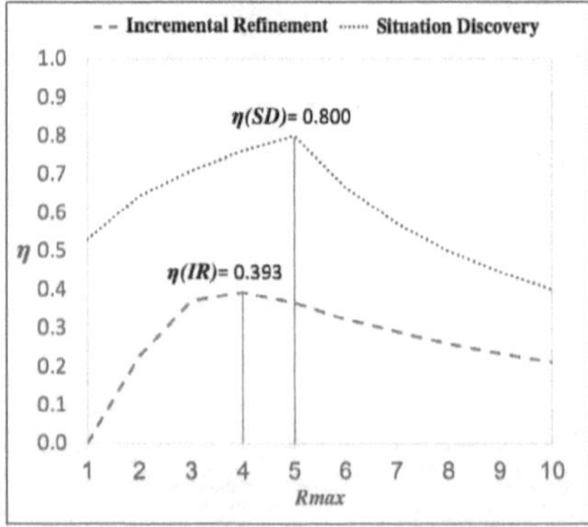

(c) The overall model performance.

Figure 5.11: Performance evaluation test: DS2, *MST*=0.3

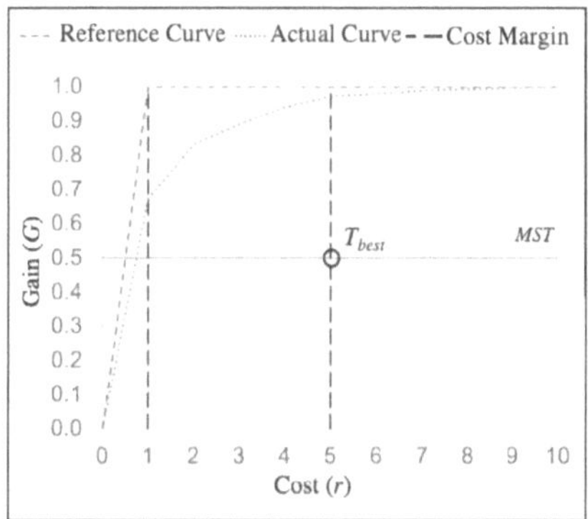

(a) *SD* mode: $\eta(SD)=0.720$ at $R_{max}=5$

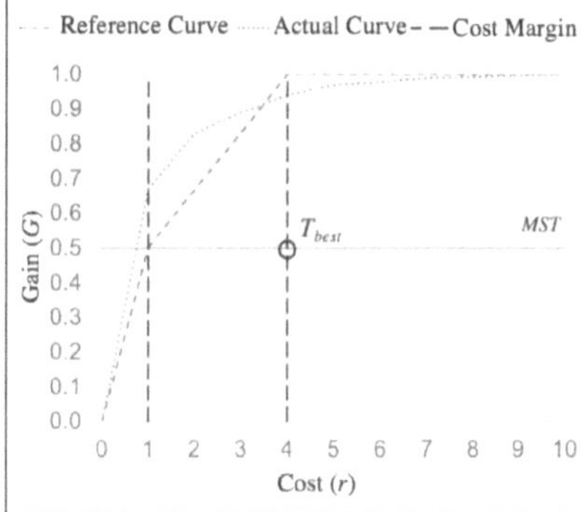

(b) *IR* mode: $\eta(IR)=0.550$ at $R_{max}=4$

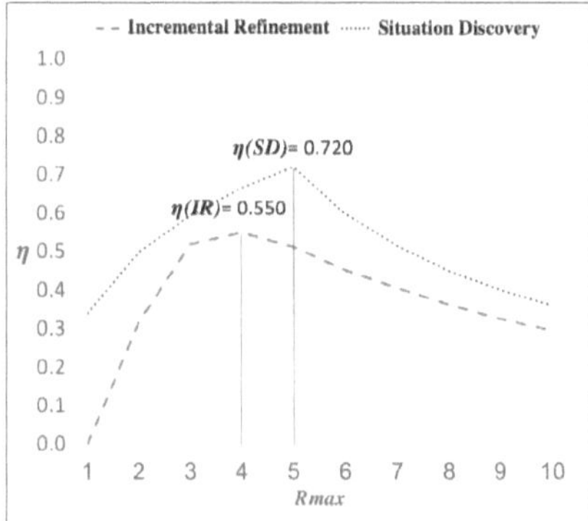

(c) The overall model performance.

Figure 5.12: Performance evaluation test: DS2, $MST=0.5$

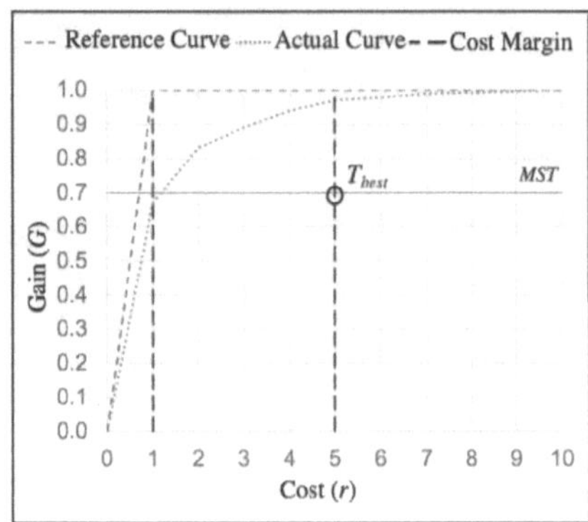

(a) *SD* mode: $\eta(SD)$=0.553 at R_{max}=5

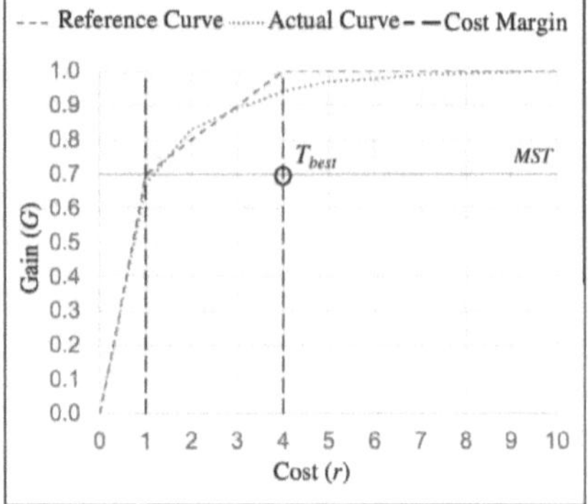

(b) *IR* mode: $\eta(IR)$=0.833 at R_{max}=4

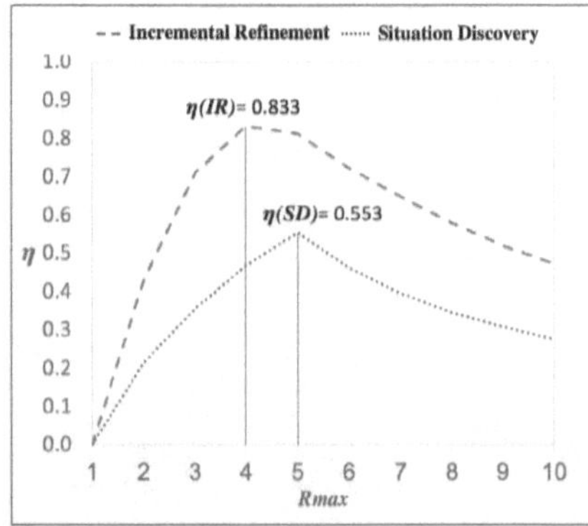

(c) The overall model performance.

Figure 5.13: Performance evaluation test: DS2, *MST*=0.7

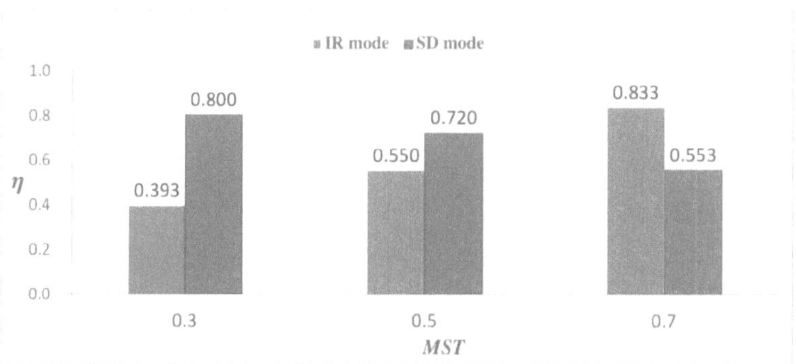

Figure 5.14: DS2: Actual Curve performance under various MST settings

5.6.2.3   Test results: Dataset (DS3)

Figures 5.15, 5.16, and 5.17 show the test results for dataset DS3 under different MST settings. Fig. 5.15 shows DS2 test results for MST=0.3, while Fig. 5.16 and 5.17 show the test results for MST=0.5 and MST =0.7, respectively. Based on the test results, the following are concluded:

• In the SD mode, the evaluation test identified T_{best} at (8, 0.3) as it delivers the highest performance with $\eta(SD) = 0.689$ (Fig. 5.15.a).

• In the IR mode, the evaluation test identified T_{best} at (6, 0.5) as it delivers the highest performance with $\eta(IR) = 0.867$ (Fig. 5.16.b).

• By observing Fig. 5.18.c, the increase in the MST threshold from 0.3 to 0.5 and 0.7 does not have a notable impact on the IR mode, whereas it negatively impacts curve performance in the SD mode.

• Overall, the given curve characteristics represent a perfect candidate for IR scenarios regardless of MST value. Besides, the given curve is not suitable for the SD scenarios due to the higher cost required to reach a higher gain (Fig. 5.17.a).

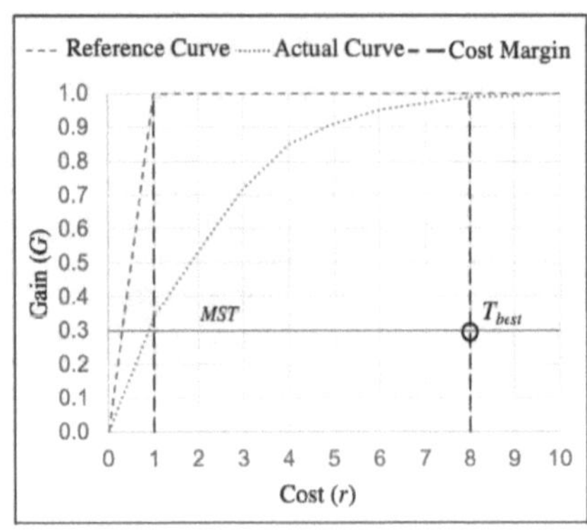

(a) *SD* mode: $\eta(SD)$=0.689 at R_{max}=8

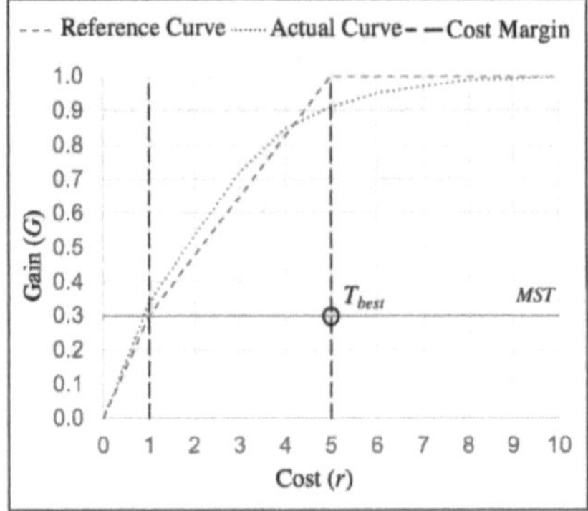

(b) *IR* mode: $\eta(IR)$=0.840 at R_{max}=5

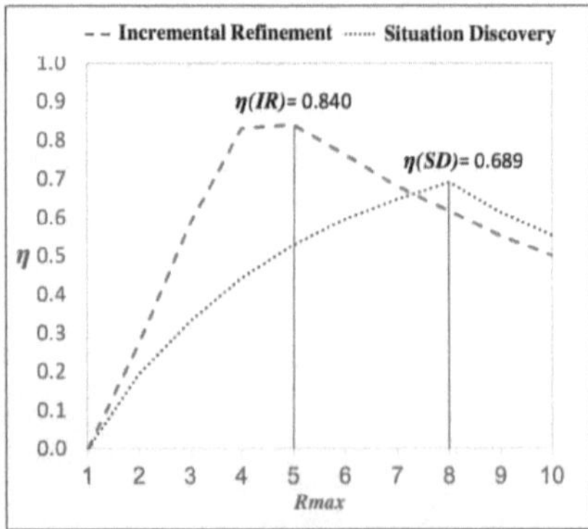

(c) The overall model performance.

Figure 5.15: Performance evaluation test: DS3, MST=0.3

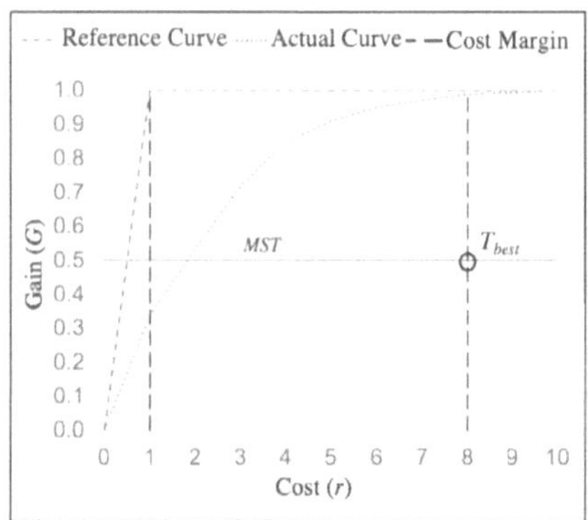

(a) *SD* mode: $\eta(SD)=0.605$ at $R_{max}=8$

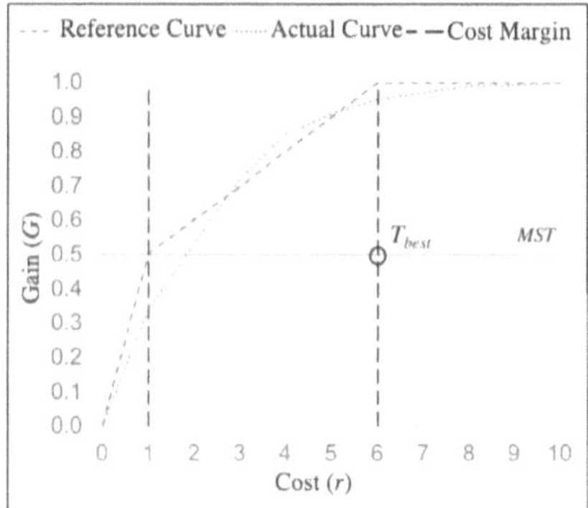

(b) *IR* mode: $\eta(IR)=0.867$ at $R_{max}=6$

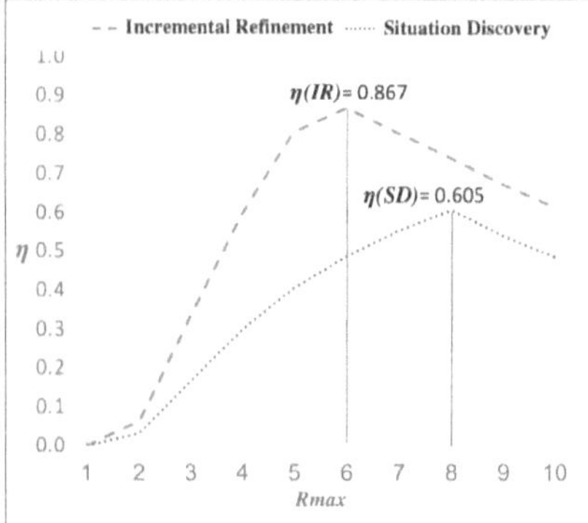

(c) The overall model performance.

Figure 5.16: Performance evaluation test: DS3, $MST=0.5$

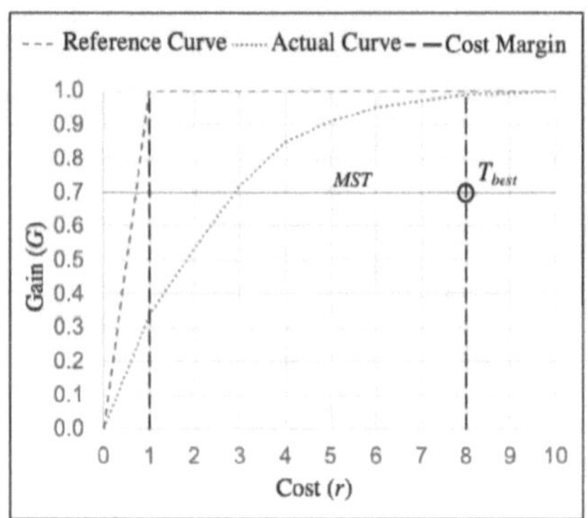

(a) SD mode: $\eta(SD)=0.496$ at $R_{max}=8$

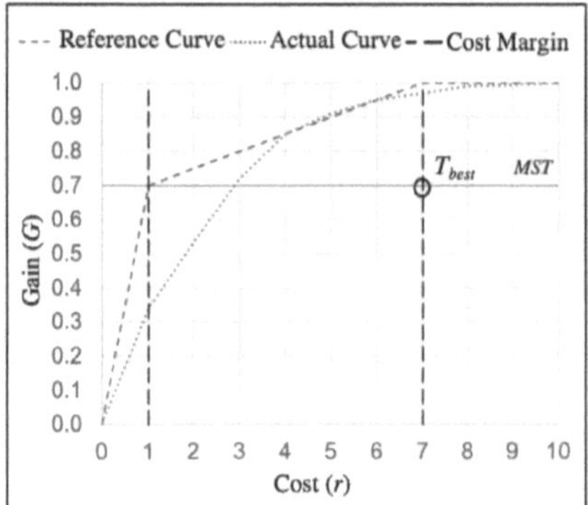

(b) IR mode: $\eta(IR)=0.838$ at $R_{max}=7$

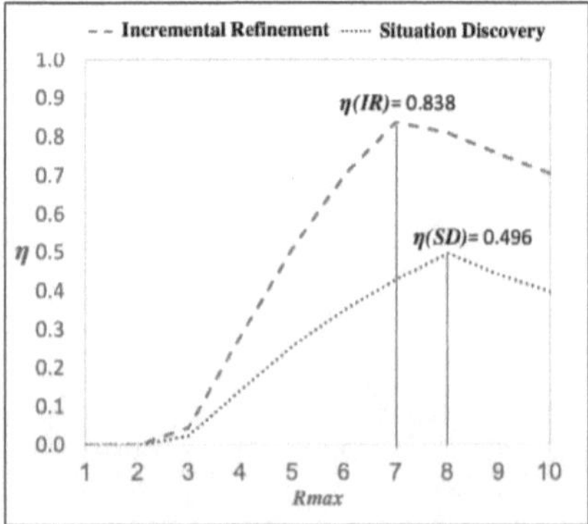

(c) The overall model performance.

Figure 5.17: Performance evaluation test: DS3, $MST=0.7$

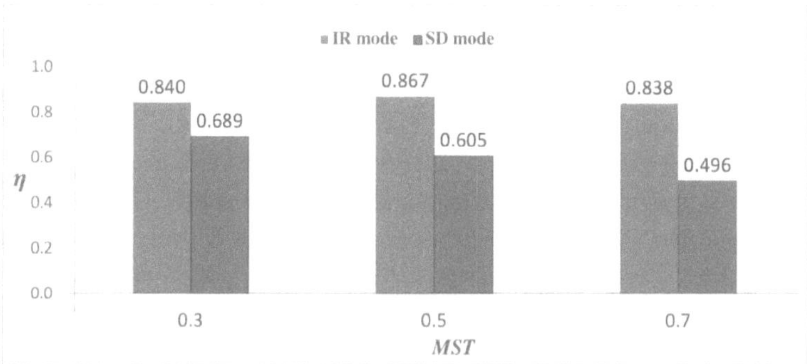

Figure 5.18: DS3: Actual Curve performance under various MST settings

5.6.3 Evaluation Test Summary

By analyzing the test results for the given datasets, the following insights are observed:

• The evaluation results of various datasets under test show the viability of the proposed approach in determining the best tradeoff to reach the maximum possible performance in a particular refinement scenario.

• Although the use of T_{best} leads to the maximum possible performance, that does not imply that the given curve is a worthy candidate for situation refinement. This can be seen in the poor performance in one test case due to the high cost required to achieve the targeted gain (DS1 in the IR mode).

• The test results show different performance trends resulting from increasing the MST threshold from 0.3 to 0.5, and 0.7. Specifically, the MST threshold increase shows a decline in the curve performance in the SD mode in all curves under test, whereas it shows an increase in performance in the IR mode. An exception to this observation is data set DS3 in the IR mode, where the given curve shows a slight increase in performance as the MST increases from 0.3 to 0.5 while it shows a slight drop in performance when changing MST from 0.5 to 0.7 (Fig. 5.18).

5.7 Chapter Summary

In this chapter, we presented a performance measure for Situation Refinement. The proposed performance measure's significance appears in determining the best trade-off between cost and gain to identify a particular SOI. The proposed performance measure uses a reference refinement curve to measure the actual curve's performance given a particular *MST* threshold setting. The proposed performance measure meets the **SRM** requirements since it accumulates actual curve coordinates for both refinement modes in a single scan as the Gradual Inclusion Process evolves. Although the results show the effectiveness of the proposed performance measure in determining the best tradeoff between cost and gain, they also show the test results show that not all refinement scenarios present a worthy candidate for situation refinement. This indicates the need to measure the viability of situation refinement in specific scenarios.

Chapter 6

book Summary

6.1   book Contribution

In this book, we presented a CEP based Situation Refinement model to re-fine the existing detection pattern towards identifying a particular *SOI*. The SRM is a tradeoff approach to address the balance between the desired level of Situation Refinement and the number of CEP rules required to support the existence of a par-ticular *SOI* as the event stream evolves.

To meet the research goals, we conducted our work according to Problem Solving Approach outlined in subsection 1.2. A summary of accomplished work at each phase is:

Phase 1: Investigating Research Problem (Chapter 1)

Various research challenges and bottlenecks that face rule learning in the event streaming environment are investigated to identify state-of-the-art **book** con-tributions.

Phase 2: Identifying Key Challenges (Chapter 3)

The motives behind knowledge enrichment in CEP and the related challenges have been investigated and classified to identify key features needed to enrich CEP rules with domain knowledge.

Phase 3: Defining Model Specific Requirements (Chapter 4)

The **SRM** concepts, roles, measures, and restrictions are defined based on var-ious types of refinements scenarios under investigation. Two algorithms have been developed toward achieving this goal, the *Gradual Inclusion Process* Algorithm and the *Summary Update* Algorithm.

Phase 4: Implementing the SRM (Chapter 4)

The **SRM** processes, components, and architecture are presented, implemented, and tested. Two software tools are used to implement SRM: the *Esper* CEP engine for Event-based Analytics [160, 162] and RStudio development platform [161]. Other tools used for various purposes related to this work include *Java Eclipse IDE* (event stream visualization) and *Excel VBA* (data conversion and presentation).

Phase 5: Evaluating the SRM outcome (Chapter 5)

The **SRM** cost-gain tradeoff measure is derived, and the model refinement outcome in both refinement modes is evaluated.

Fig. 6.1 illustrates the dissertation's contribution summary.

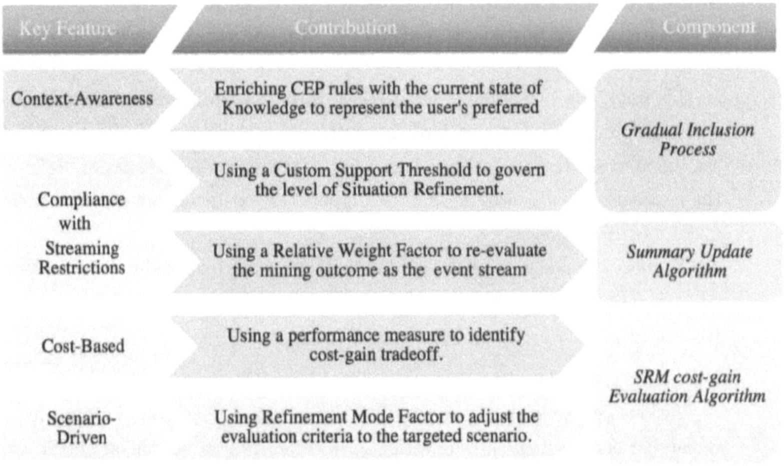

Figure 6.1: Research Contribution Summary

6.2 Conclusions

In this book, we addressed various challenges that face CEP systems toward achieving a higher level of Situational Awareness. While the existing approaches fo-cus on enhancing the level of expressiveness of knowledge presentation or reducing the number of detection rules to optimize CEP query processing, we used a different path toward mitigating this issue by a tradeoff approach that takes into account both sides of the problem. In particular, we aimed to answer the primary research question related to our work:

Q: "How to gain the desired level of situation refinement while maintaining the minimum number of CEP rules?"

Based on the findings and results gained throughout this work, the following conclusions are made:

1. In contrast to existing Situation Refinement approaches where one or at most two key features are under focus, four key features have been under focus in this work; Context-Awareness, Compliance with event streaming restrictions, Cost-Based Situation Refinement, and Scenario-Driven Situation-Refinement. As a result, three key challenges have been successfully addressed in this work: the tradeoff between pattern complexity and pattern coverage, the compliance with event streaming constraints, and the *SOI* selectivity based on refinement's cost and gain.

2. Contrary to existing fine-tuning methods where two factors are taken into consideration (available resources vs. time), we show the importance of adopting a refinement evaluation criterion that takes an additional factor into account, i.e., the situation complexity. The importance of such a factor is justified since the task of situation comprehension depends mostly on domain experts. Therefore, using the minimum number of detection rules is essential for Situational Awareness.

3. In this work, we show the importance of incorporating the unique business requirements of the refinement scenario toward the optimal refinement outcome. These requirements have been investigated, identified, and presented in our work in [10], [28]. Based on that, we identified two types of refinement scenarios with unique outcome curve characteristics (*Incremental Refinement, Situation*

Discovery). In addition, we introduced a single scan measure of performance that is sensitive to both refinement modes.

4. In this work, we introduced the notion of Refinement Zone (*RZ*) to evaluate the refinement scenario in a single scan while adapting actual refinement boundaries for the given mode. The **SRM** performance evaluation results show that the size of *RZ* and the ratio between its boundaries can be used to reveal additional features about the *SOI* under evaluation. For example, a larger *RZ* size with H/W ratio close to 1.0 indicates that the given *SOI* is a strong candidate for *Incremental Refinement*. Furthermore, the H/W ratio that is greater than 1.0 indicates that the given *SOI* performs better in the *Situation Discovery* mode, whereas the H/W ratio smaller than 1.0 means that the given *SOI* performs better in *Incremental Refinement* mode.

5. Although the **SRM** follows the event streaming guidelines concerning delivering the mined items instantly upon the user request, the test results presented in Section 4.6.4 show stability in the refinement curve as the situation discovery continues for an extended period. Such stability indicates that recent trends will have a minimal impact on the overall situation discovery. Therefore, establishing maximum discovery time must be considered in any future enhancement. One possible approach to mitigate this issue is by further investigating the overall trend of the refinement curve to determine such a limit. Another approach that can be used for more extended discovery periods is by applying the *time-fading* model to provide the most recent part of the event stream with more weight than the past summary.

Appendix A

Software Tools Used in this Research

A.1   Design and Programming Tools

1. **RStudio 1.3.9** Integrated Development Environment (IDE) for R

   - Implement R code for various algorithms.

2. **Eclipse 4.5.0** Integrated Development Environment [IDE] – Java 1.8 plug in:

   - Create various solution functions & modules.

3. **Esper 5.3.0** Complex event processing [CEP] engine.

   - Create CEP pattern template.
   - Detect and analyse complex event patterns.

4. **Apache Maven 3.3.9** software project management (POM).

   - Manage a project's build, reporting and documentation.

A.2   Data Implementation and Visualization Tools

1. **R 3.2.4** programming language and data mining tool:

   - Perform various data analytics statistical tasks.

2. **MySQL 5.7** Relational Database Management System (RDBMS)

   - Store and retrieve knowledge base data.

   - Store pattern rules.

3. **Excel VBA** Microsoft Excel Visual Basic Macro

   - Data presentation and visualization.

   - Dataset and results migration tasks.

A.3   Add-on Libraries

1. *SPARQL* Event Query Language Library.

2. *Espertech* Esper Complex Event Processing Library.

3. *SML* Semantic Measures Library.

Appendix B

The SRM Test Suite

```
################################################################################
#                     This R code is used   for the following:                #
#                     1- Implement the SRM test modules:                       #
#                          - The Gradual Inclusion module                      #
#                          - Summary Update module                             #
#                     2- Import and fire event data                           #
#                     3- Export test results into CSV format  for result visualization#
################################################################################
#------------------------------------------------------------------------------
#------------------------------------------------------------------------------
#------------------------------- MAIN Program ---------------------------------
#------------------------------------------------------------------------------
#------------------------------------------------------------------------------

library(lubridate)
dev.off(dev.list()["RStudioGD"])
cat("\014")
setwd("/Users/alaaalakari/Desktop/work/data/R")
SRM<- new.env()

#-------------- Defining Parameters --------------

assign('a',  NULL,envir=SRM)
assign('b',NULL,envir=SRM)
assign('va',  NULL,envir=SRM)
assign('vb',NULL,envir=SRM)
assign('w_size',NULL,envir=SRM)
assign('no_of_rules',NULL,envir=SRM)
assign('Max_level_complexity',0,envir=SRM)
assign('sample',0,envir=SRM)
assign('MST',0,envir=SRM)
assign('total_window',0,envir=SRM)
assign('current_weight',0,envir=SRM)
assign('rules',NULL,envir=SRM)
assign('rules_weight',NULL,envir=SRM)
```

```
define()  # ---    USER DEFINED PARAMETERS ---------------

#====   Initialize  events table ===========
a <<- matrix(SRM$no_of_rules) # event array
b <<- matrix(SRM$no_of_rules) # counting array
rules<<- matrix(SRM$no_of_rules,10) # rules
rules_weight<<- matrix(SRM$no_of_rules,10) # rules weight


#======= READ input dataset   ===================================

data <- read.csv("dataset.csv",  strip.white = TRUE) # read stream input

#================================================================

reset_vars()  # reset all varibles
for (q in 1:SRM$no_of_rules){SRM$rules[q]<-"";SRM$rules_weight[q]<-0.0}
#-------------------------------------------------------
for(k in 1:1)   # k represents number of Folds
{
for (q in 1:SRM$no_of_rules){SRM$rules[q]<-"";SRM$rules_weight[q]<-0.0}
SRM$total_window<-0   # total window size
SRM$current_weight<-0
i<-1    # i represents starting row in dataset
i=(SRM$sample*(k-1))+1
t1<-data[i,3]
m_row<-nrow(data)

while(data[i,3]<=SRM$sample)
{
reset_vars()  # reset all varibles
counter<-1
win<-0

# ----- start of new time window
while(counter<=SRM$w_size) # counter represents no of second per window
{
t2=t1
while(t1==t2)
{
if(i==m_row){break} # end of stream detected
modify(SRM,i,data[i,1,1])
i<-i+1
win<-win+1
t1<-data[i,3]
}

if(i==m_row){break}
t1<-data[i,3]
counter<-counter+1  #  increment within time window
}
if(i==m_row){break}
```

```r
# ........  End of time window .............


SRM$total_window<-SRM$total_window+win  # calculating total window
SRM$current_weight<-win/SRM$total_window  # calculating current window weight ( W =
    win/total   )


cat("no of events per window:",win,"\n")
cat("total window size so far:",SRM$total_window,"\n")
cat("current window weight:",SRM$current_weight,"\n")
cat("past weight:",(1-SRM$current_weight),"\n")
cat("====:",i,"\n")
migrate(SRM,i)
}


#####################################################
####### USER DEFINED functions #######################
#####################################################

# USER DEFINED PARAMETERS  -----------------------
define <-function()
{
SRM$w_size<-2 # counter represents no of seconds per window. i.e =3 seconds
SRM$no_of_rules<-10 # no of rules --> buffer size
SRM$MST<-0.57   # minmum support threshold
SRM$Max_level_complexity<-3 # maximum level of rule complexity
SRM$sample <-30  # sample size (in seconds)

SRM$results<- as.data.frame(matrix(0,nrow=5,ncol=4))  # Data Frame for recording
    results
SRM$resultsIndex<-0  # row number in results DF
}
#----------------------------------------------

#  Varible reset
reset_vars <- function()
{
#=============================================================
for(q in 1:SRM$no_of_rules)
{
SRM$a[q]<-""
SRM$b[q]<-0
SRM$acc_A[q]<-""
SRM$acc_B[q]<-0.0
}
}
# ---------  Printing
printOut <- function()
{
sum1<-0
sum2<-0
for (q in 1:SRM$no_of_rules){sum1<-sum1+SRM$rules_weight[q]}
```

```
for (q in 1:SRM$no_of_rules){SRM$rules_weight[q]<-SRM$rules_weight[q]/sum1;sum2<-sum2
    +SRM$rules_weight[q];
if(SRM$rules_weight[q]>0.0001)
{
cat(SRM$rules[q],SRM$rules_weight[q],"\n")

# ===== Writing results in results DF
SRM$resultsIndex=SRM$resultsIndex+1    # increase row number in DF
SRM$results[SRM$resultsIndex,1]=k  # Fold number
SRM$results[SRM$resultsIndex,2]=SRM$rules[q]     # Rule
SRM$results[SRM$resultsIndex,3]=SRM$rules_weight[q]  # Rule weight
#==   End writing

}
}
}
insertRow <- function(existingDF, newrow, r) {
existingDF[seq(r+1,nrow(existingDF)+1),] <- existingDF[seq(r,nrow(existingDF)),]
existingDF[r,] <- newrow
existingDF
}


#=============
migrate <- function(SRM,i)
{

SRM$va<-cbind(SRM$va,as.vector(SRM$a))
SRM$vb<-cbind(SRM$vb,as.vector(SRM$b))
update_rules(SRM,i)
#    for (q in 1:SRM$no_of_rules){SRM$a[q]<-" ";SRM$b[q]<-0}
}


update_rules<-function(SRM,i)
{
rule_sum<-0
select_rules(SRM,i)
}
select_rules<-function(SRM,i)
{
cr<-""     # current rule
cr_weight<-0.0    # current weight for the rule
cat("current window:",win,"   current weight:",SRM$current_weight,"\n")
for (q in 1:SRM$Max_level_complexity)
{
cat("list of rules:",SRM$a[q],"   events:",SRM$b[q],"\n")
s<-1
flag<-0
while(s<=SRM$no_of_rules )
{
if(SRM$a[q]==SRM$rules[s])
{
```

```
SRM$rules_weight[s]<- SRM$rules_weight[s]+(SRM$b[q]*SRM$current_weight) # weight
    rebalancing (past & current)
s<-SRM$no_of_rules
flag<-1
}
if((SRM$rules[s]=="") && (flag==0))
{
SRM$rules[s]<-SRM$a[q];
SRM$rules_weight[s]<- SRM$b[q]*SRM$current_weight
s<-SRM$no_of_rules
}
s<-s+1
flag<-0
}
}
}


#============
modify <- function(SRM,i,m) {
j <- 1
flag <- 0
s <- " ";

# Updating Event Table
while (j < SRM$no_of_rules)
{
if (SRM$a[j] == m)
{
flag <- 1
SRM$b[j] <- SRM$b[j] + 1
}
if (SRM$a[j] == "")
{
s <- toString(data[i, 1, 1])
SRM$a[j] <- s
SRM$b[j] <- SRM$b[j] + 1
flag <- 2
}

if (flag > 0)   # means an update exist
{
#============CHECK SWAP ===============================
if (SRM$b[j] > SRM$b[j - 1] && j > 1)
{
t <- SRM$b[j]
SRM$b[j] <- SRM$b[j - 1]
SRM$b[j - 1] <- t
t1 <- SRM$a[j]
SRM$a[j] <- SRM$a[j - 1]
SRM$a[j - 1] <- t1
}
#========== END SWAP ===============================
break
```

```
}
j <- j + 1
}#- end of events table update
}
eventDelay<-function(x)
{
m<-0
while(m<x)
{
t1<-as.integer(second(Sys.time()))
m=m+1
}
}

########## Prinintg results on the screen
printOut()
write.csv(SRM$results,'results.csv')
}
#=====================================
```

Appendix C

Cost-Gain Evaluation Algorithm

```r
####################################################################
# This R code is used to implemnt cost-gain evaluation algorithm  #
####################################################################

# Defining parameters

library(lubridate)
dev.off(dev.list()["RStudioGD"])
cat("\014")
setwd("/Users/alaaalakari/Desktop/work/data/R")

cost<- new.env()
assign('i', 0,envir=cost)
assign('Cri', 0,envir=cost)
assign('a', 0,envir=cost)
assign('b', 0,envir=cost)
assign('x', 0,envir=cost)
assign('a_i', 0.0 ,envir=cost)
assign('MST',0.0 ,envir=cost) # Minimum Support Threshold
assign('Rmax',0,envir=cost) # Level of complexity
assign('alpha',0,envir=cost)
assign('beta',0.0,envir=cost)
assign('Cm_i',0.0,envir=cost)
assign('Cm',0.0 ,envir=cost)
assign('M',0,envir=cost)
assign('m_IR',0.0,envir=cost)
assign('m_SD',0.0 ,envir=cost)
assign('m_SD_max',0.0 ,envir=cost)
assign('m_IR_max',0.0 ,envir=cost)
assign('Tbest_IR',0,envir=cost)
assign('Tbest_SD',0,envir=cost)

cost$results<- as.data.frame(matrix(0,nrow=10,ncol=2))  # Data Frame for
    recording results
ds <<- matrix(10) #
```

```r
#========= Acquire  Dataset under test
cost$data <- read.csv("ds1.csv",  strip.white = TRUE)
#cost$data <- read.csv("ds2.csv",  strip.white = TRUE)
#cost$data <- read.csv("ds3.csv",  strip.white = TRUE)

for(q in 0:9)
{
ds[q]<-cost$data[q,1];
cat("BMW:",ds[q],"\u"),
}

MST<-0.4;
Rmax<-0.3;

alpha<-Rmax-1;
beta<-1-MST;
b_i<-0.0;
i<-1;
b<-0;
Cm<-0;
Cm_i<-0;
m_SD_max<-0;
m_IR_max<-0;
Tbest_IR<-0;
Tbest_SD<-0;

while(i<=Rmax)
{
cost$b_i<-0;
Cri<-ds[i];


# Calculate  additional gain  in IR mode
a_i<-0;
a_i<-beta/alpha*(i-1);
Cm_i<-(Cri-MST);

if (Cm_i<0){Cm_i<-cost$Cm_i*-1.0}
if(Cm_i<=0)
{
cost$b_i<-beta;
cat("b1:",cost$b_i,"\n");
}
else
{
cost$b_i<-(cost$Cm_i-a_i);
}

# Accumulate  the absolute difference in IR mode
b<-b+b_i;

# Accumulate the additional gain in SD mode
```

```
Cm <- Cm +(( Cm_i + abs ( Cm_i ))/2) ;

#Calculate}  performance for both modes

m_IR <- -1 -b/ c ;
if ( Rmax ==1) { m_IR =0;}

m_SD <- Cm /( beta * Rmax ) ;
if  ( m_IR  >  m_IR_max )
{
m_IR_max <- m_IR ;
Tbest_IR <- i ;
}

if  ( m_SD  >  m_SD_max )
{
m_SD_max <- m_SD ;
Tbest_SD <- i ;
}
i <- i +1;
}
cost$results [i-1,1] <- m_SD_max ;  cost$results [i-1,2] <- m_IR_max ;
write.csv ( cost$results , 'costResults.csv ')
```